AUTOMATED DIAGNOSIS SYSTEM
FOR MALARIA DETECTION AND ANALYSIS

SANJAY NAG

TABLE OF CONTENTS

TABLE OF CONTENTS V
LIST OF FIGURES VIII
LIST OF TABLES XIII
LIST OF ACRONYMS USED XV
CHAPTER 1 1
INTRODUCTION 1
   1.1. OVERVIEW 1
   1.2. DIGITAL PATHOLOGY 2
   1.3. MALARIA DISEASE 3
      1.3.1. Socio-economic Impact 4
      1.3.2. Malaria Statistics 5
      1.3.3. Biological Aspect of Malaria Disease 5
         1.3.3.1. Plasmodium Lifecycle 8
         1.3.3.2. Composition of Blood 9
      1.3.4. Diagnosis of Malaria 10
         1.3.4.1. Clinical Diagnosis of Malaria 10
         1.3.4.2. Laboratory Diagnosis of Malaria 11
         1.3.4.3. Molecular Diagnosis of Malaria 15
         1.3.4.4. Virtual Microscopy 17
   1.4. RESEARCH OBJECTIVES 17
   1.5. CONTRIBUTION TO THE KNOWLEDGE DOMAIN 18
   1.6. ORGANIZATION OF THE THESIS 19
CHAPTER 2 21
LITERATURE REVIEW 21
   2.1. OVERVIEW 21
   2.2. REVIEW MATERIALS AND METHODS 22
      2.2.1. Article Crawling 22
      2.2.2. Selection of Articles 22
   2.3. REVIEW METHODS 25
   2.4. DIGITAL MICROSCOPY BASED CAD METHODS 27
      2.4.1. Pre-processing 28
      2.4.2. Rule Based Segmentation 29
      2.4.3. Rule Based Parasite Detection 33

  2.4.4. Learning Based Segmentation and Parasite Detection 34
 2.5. RESEARCH GAP ANALYSIS.. 38
 2.6. SUMMARY 40

CHAPTER 3 42

OVERVIEW OF DATA MODEL..42

 3.1. BLOCK REPRESENTATION OF THE DATA MODEL 42
 3.2. A BRIEF OUTLINE OF THE PROPOSED MODEL 43

CHAPTER 4 46

DATA DESCRIPTOR..46

 4.1. OVERVIEW 46
 4.2. DATASET PREPARATION & SAMPLING... 46
  4.2.1. Initial Screening.. 46
  4.2.2. Detection and Classification Method 47
   4.2.2.1. Dataset #1 47
   4.2.2.2. Dataset #2 47
   4.2.2.3. Dataset #3 47
  4.2.3. Dataset Annotation.. 50
  4.2.4. Dataset Standardization 50
 4.3. SUMMARY 50

CHAPTER 5 51

IMAGE PRE-PROCESSING 51

 5.1. OVERVIEW 51
 5.2. ILLUMINATION CORRECTION 52
 5.3. FOR INITIAL SCREENING 54
 5.4. MAIN PHASE 56
  5.4.1. Background Separation... 56
  5.4.2. Reference Scale Artefact Removal 57
  5.4.3. Non-uniform Background Pixel Removal 57
  5.4.4. Image Recoloring 59
  5.4.5. Declumping 63
  5.4.6. Parasitaemia estimation 69
 5.5. SUMMARY 70

CHAPTER 6 71

FEATURE SELECTION/EXTRACTION 71

 6.1. OVERVIEW 71
 6.2. INITIAL SCREENING PHASE 71
 6.3. MAIN PHASE 79
 6.4. SUMMARY 84

CHAPTER 7 85

CLASSIFICATION 85

 7.1. OVERVIEW 85
 7.2. INITIAL SCREENING 85
 7.3. MAIN PHASE 87
  7.3.1. Singular Classifiers 87
   7.3.1.1. K-NN Classifier...87
   7.3.1.2. Naïve Bayes Classifier..87

| 7.3.1.3 Support Vector Machine 87
      7.3.2. Ensemble Classifiers 88
   7.4. SUMMARY 90

CHAPTER 8 92

RESULTS & DISCUSSION 92

   8.1. OVERVIEW 92
   8.2. INITIAL SCREENING PHASE 92
      8.2.1. Performance Metrics and Discussion 92
      8.2.2. Comparative Study 95
   8.3. MAIN PHASE 97
      8.3.1. Performance Metrics and Discussion 97
        8.3.1.1. Detailed Stage/Specie based Classification (Cellular Level) 107
        8.3.1.2. Image-Level Classification of Infected and Normal Images 112
      8.3.2. Comparative Study 113
   8.4. SUMMARY 119

CHAPTER 9 120

LIMITATION 120

   9.1. OVERVIEW 120
   9.2. TOO MANY HUGE BLOOD CELL CLUMPS 120
   9.3. LOW RESOLUTION/POOR QUALITY IMAGE 122
   9.4. IMAGE WHERE CYTOPLASM IS HEAVILY STAINED 123
   9.5. SUMMARY 124

CHAPTER 10 125

COST ANALYSIS OF PROPOSED SYSTEM 125

   10.1. OVERVIEW 125
   10.2. PRESENT SYSTEM 125
   10.3. PROPOSED SYSTEM 125
   10.4. COMPARISON OF CAD SYSTEM (PROPOSED) WITH MANUAL SYSTEM (EXISTING) 126
   10.5. SUMMARY 127

CHAPTER 11 128

CONCLUSION 128

   11.1. RESEARCH OUTCOME 128
   11.2. FUTURE WORK 129

REFERENCES 131

LIST OF RELEVANT PUBLICATIONS 142

   JOURNALS 142
   BOOK CHAPTER 142

LIST OF FIGURES

FIGURE 1. A COMPARISON OF THE DIFFERENT LIFECYCLE FORMS OF PLASMODIUM GENUS. FIGURE (A)-(C) ARE TROPHOZOITE OF P. VIVAX, MATURE SCHIZONT OF P. VIVAX, GAMETOCYTE OF P. VIVAX AND (D)- (F) TROPHOZOITE OF P. FALCIPARUM, MATURE SCHIZONT OF P. FALCIPARUM, GAMETOCYTE OF P. FALCIPARUM. THE IMAGES OBTAINED FROM CDC, DPDX- MALARIA IMAGE LIBRARY [24] 8

FIGURE 2. THE LIFE CYCLE OF PLASMODIUM GENUS AND ITS DEVELOPMENT IN HUMAN AND MOSQUITO HOSTS. 9

FIGURE 3. THE FIGURE SHOWS FIVE DIFFERENT TYPES OF WBC WITH NEUTROPHIL AT THE LEFTMOST POSITION FOLLOWED BY MONOCYTE, BASOPHIL, LYMPHOCYTE AND EOSINOPHIL AT THE EXTREME RIGHT. THIS RARE IMAGE WAS RECORDED BY A HAEMATOLOGIST WHILE SCANNING THROUGH SMEAR SLIDES............ 10

FIGURE 4. TOTAL OF 291 RESEARCH ARTICLES ON MALARIA AND RELATED MEDICAL TEXTS WAS INVESTIGATED 23

FIGURE 5. DISTRIBUTION OF THE 237 RESEARCH ARTICLES UNDER CONSIDERATION FOR THE REVIEW WORK 24

FIGURE 6. DISTRIBUTION OF TOTAL 172 PAPERS INCLUDING JOURNAL, CONFERENCE PROCEEDINGS AND BOOK CHAPTERSWITH REFERENCE TO THE PUBLICATION YEAR .. 24

FIGURE 7. GENERALIZED SCHEME FOR DETECTION OF MALARIA PARASITE 27

FIGURE 8. GAP ANALYSIS AS AN OUTCOME OF THOROUGH LITERATURE REVIEW. THE POOSIBLE WAYS TO MEND THE GAPS PRESENT IN THE CURRENT SYSTEM............ 40

FIGURE 9. BLOCK DIAGRAM OF THE PROPOSED MODEL 42

FIGURE 10. FLOWCHART OF THE PARASITE DETECTION SYSTEM............ 45

FIGURE 11. SAMPLE DATASET AT 1000X (A) MAMIC IMAGE SHOWING MATURE P. VIVAX SCHIZONT, (B) MAMIC IMAGE SHOWING MULTIPLE INFECTION OF P. FALCIPARUM RINGS, (C) HOSPITAL SUPPLIED SLIDE IMAGE SHOWING P. VIVAX GAMETOCYTE AND (D) HOSPITAL SUPPLIED SLIDE IMAGE SHOWING P. FALCIPARUM SCHIZONT............ 50

FIGURE 12. (A) AND (B): TWO SAMPLE IMAGES FROM THE DATASET (FOREGROUND EXTRACTED IMAGE) SHOWING THE INTER IMAGE COLOUR VARIATIONS, EACH HAVING A PARTICULAR SET OF UNIQUE COLOURS WHICH ARE STATISTICALLY SIGNIFICANTLY DIFFERENT.(C) AN ENLARGED VIEW OF A RED BLOOD CORPUSCLE ELUCIDATING INTRA-CELLULAR COLOUR VARIATION.IMAGE SHOWING INTRA AND INTER DATASET VARIATIOS 51

FIGURE 13. (A) AN IMAGE WHOSE LUMINANCE HAS BEEN DEGRADED ON PURPOSE TO TEST WHETHER THE ALGORITHM WORKS ON OTHER IMAGE DATASETS WITH BAD LUMINANCE.[LUMINANCE DISTRIBUTION SPECIFICATION PROVIDED IN FIGURE 14].(B) A REFERENCE IMAGE FROM THE MAMIC DATABASE TO IMPROVE THE ILLUMINATION OF IMAGE (C) LUMINANCE IMPROVED IMAGE............ 52

FIGURE 14. REPRESENTS THE HISTOGRAM OF THE LUMINANCE MARRED IMAGE OF THE MAMIC DATABASE (ORIGINAL IMAGE MAMIC_57503.BMP) ALONG WITH THE ORIGINAL SPECIFICATIONS FOR THE SAME. ADDITIONALLY, THE SPECIFICATION FOR THE LUMINANCE CHANNEL OF THE REFERENCE IMAGE AND THE FINAL ILLUMINATED

IMAGE ALONG WITH THE LUMINANCE CHANNEL DISTRIBUTION SPECIFICATION HAVE ALSO BEEN DOCUMENTED IN THE IMAGE. 53

FIGURE 15. BLOCK DIAGRAM DEPICTING THE PROCESS OF SEGEMNTATION OF THE IMAGE TO OBTAIN THE FOREGROUND REGION. 54

FIGURE 16. (A) THIN BLOOD SMEAR IMAGE, (B) CLUSTER 1 CONSISTING OF RED BLOOD CELLS, (C) CLUSTER 2 CONSISTING OF WHITE BLOOD CELLS, PLATELETS AND MALARIAL PARASITE. 55

FIGURE 17. DISC SHAPED EROSION WAS PERFORMED TO REMOVE TRACE OF WBC OUTLINE WITHIN THE RBC CLUSTER 56

FIGURE 18. SAMPLE IMAGE FROM MAMIC DATABASE, (A) INPUT IMAGE (RGB) WITH BACKGROUND, (B) THE IMAGE IN RGB COLOUR SPACE. (C) THE FOREGROUND IDENTIFICATION THAT WAS USED AS REFERENCE POINT (GROUND TRUTH) FOR EVALUATION OF THE PERFORMANCE OF EXTRACTION ALGORITHMS 57

FIGURE 19. OUTPUT IMAGE FOR AUTOMATIC FOREGROUND EXTRACTION (A) USING 3-MEANS CLUSTERING (B) OUTPUT OBTAINED FOR 3 THRESHOLD BASED ZACK'S CLUSTERING ALGORITHM BASED ON THE 3 MOST PROMINENT PEAKS IN THE INDEXED IMAGE HISTOGRAM. 58

FIGURE 20. (A) AND (C) REPRESENT SAMPLE IMAGES FROM MAMIC DATASET (B) AND (D) DOCUMENT THE EXTRACTED FOREGROUND CELLULAR PARTICLES AFTER SUPPRESSION OF NON-UNIFORM BACKGROUND. AS PER WHO GUIDELINES FOR MALARIA MICROSCOPY, THE CELLS IN THE BORDER REGION WERE REMOVED. 59

FIGURE 21. (A) AND (C) THE FOREGROUND EXTRACTED SAMPLE IMAGES CONVERTED TO YC_BC_R COLOUR SPACE; (B) AND (D) REPRESENT THE RECOLOURED DOWN SAMPLED IMAGE [REPRESENTED WITH 4 COLOURS IN RGB COLOUR SPACE (THE FOURTH IS BLACK)] 60

FIGURE 22. (A) THE RECOLOURED SAMPLE IMAGE WHERE ALL RED BLOOD CORPUSCLES ARE NORMAL (B) THE ENLARGED VIEW OF A NORMAL CELL COMPONENT IN THE IMAGE REPRESENTING THE POINT SELECTION PROCESS ADOPTED FOR ANALYSING EACH CONNECTED COMPONENT (C) THE RECOLOURED IMAGE PART FROM ANOTHER IMAGE IN THE DATASET REPRESENTING PRESENCE OF MALARIA IN AN ERYTHROCYTE.[THE GREEN REGION REPRESENTS THE INFECTION] (D) REPRESENTATION OF A WHITE AND RED BLOOD CELL CLUSTER FROM THE SAME 62

FIGURE 23. (A) INPUT IMAGE FROM MAMIC DATASET CONTAINING NORMAL IMAGE (B) THE OUTPUT IMAGE IDENTIFYING ONLY THE PLATELET AS A SUSPECT REGION [PLATELETS ELIMINATED BASED ON TUKEY'S HINGE](C) INPUT IMAGE CONTAINING A PARASITE WITHIN A RED BLOOD CELL AND A WHITE BLOOD CELL (D) THE OUTPUT IMAGE SHOWING THE SUSPECT REGION. 63

FIGURE 24. (A) THE ORIGINAL IMAGE (B) THE BINARY MASK IN (A) IS REPRESENTED ALONG WITH THE AREA OF THE CLUMPED CELLS. THE MARKED OUT CLUMPS IN THE IMAGE ARE NUMBERED IN ORDER TO MATCH THE SAME WITH THE CORRESPONDING DE-CLUMPED VERSIONS. (C) IMAGE FROM THE DATASET CONSISTING OF A MIXED CLUSTER OF RBC AND WHITE BLOOD CELL/S. (D) IMAGE IN (B) CONSISTS OF ONLY ONE CLUSTER. DE-CLUMP OF THE MIXED WHITE AND RED BLOOD CELL/S. 65

FIGURE 25. THE RECONSTRUCTED DE-CLUMPED BINARY IMAGE MASK FOR THE IMAGE IN FIGURE 24A 66

FIGURE 26. (A) A SAMPLE IMAGE CONTAINING INFECTED RBC AND TWO WBC CLUMPED WITH OTHER INFECTED/NON-INFECTED RED BLOOD CELLS. (B) THE BINARY MASK OF THE DE-CLUMPED AND RECONSTRUCTED IMAGE WITH THE WHITE BLOOD CELLS MARKED WITH 'W', THE RED BLOOD CORPUSCLES MARKED WITH 'R' AND THE ELEMENTS TO BE DISCARDED MARKED WITH D. 66

FIGURE 27. REPRESENTATION IMAGES THAT WERE USED FOR PIXEL POSITION BASED MATCHING IN ORDER TO DIFFERENTIATE THE NORMAL (I.E. WHITE BLOOD CELL NUCLEUS) FROM THE MALARIA PARASITE IN THE SUSPECTED PROBLEM AREA CLUSTER (A) ORIGINAL DIGITIZED THIN BLOOD SMEAR SAMPLE IMAGE ; (B) THE DE-CLUMPED BINARY MASK DEVELOPED; (C) THE 4 COLOUR CODED IMAGE IN THE YC_BC_R COLOUR SPACE; (D) THE SUSPECTED PROBLEM CLUSTER REPRESENTED IN BINARY 68

FIGURE 28. A BRIEF PICTORIAL REPRESENTATION OF THE ALGORITHM USED FOR RE-CLUMPING OF A FALSELY DE-CLUMPED WHITE BLOOD CELL. (A) THE CLUMPS (WHITE BLOOD CELL CLUMPS), (B) FOR THE CLUMP THE DE-CLUMPING MARKERS ARE REPRESENTED IN WHITE WHILE EACH OF THE LINES (I.E. THE LINE JOINING THE CENTER OF EACH SEPARATE CLUMP TO THE CENTROID OF THE CLUMPS TAKEN AS A WHOLE) HAVE BEEN REPRESENTED IN BLUE AND RED RESPECTIVELY. (C) IF ANY OF THE LINES INTERSECT THE DE-CLUMPING MARKERS, THE MARKER IS REMOVED FROM THE RECONSTRUCTED IMAGE AS REPRESENTED IN THE GIVEN FIGURE................................ 69

FIGURE 29. (A) THE RE-COLOURED IMAGE SHOWING TWO SEPARATE CLUSTERS IN A SINGLE FRAME. THE IMAGE ALSO INCLUDES ERYTHROCYTE ENUMERATION (B) ELIMINATION OF WHITE BLOOD CELL NUCLEUS RESULTS IN THE ACTUAL MARKING OUT OF THE MALARIA PARASITE INFECTION. 70

FIGURE 30. RED BLOOD CELLS WITH SUSPECTED MALARIAL PARASITE WERE DISTINCTLY MARKED OUT FROM THE RBC CLUSTER 71

FIGURE 31. THE SUSPECTED MALARIAL PARASITE PRESENT WITHIN RBC IS MARKED IN RED. THE EDGES OF THE ERYTHROCYTE CELLS THAT ARE DEEMED OR SUSPECTED TO CONTAIN SUSPECTED MALARIAL PARASITE ARE MARKED OUT 72

FIGURE 32. CLUSTER 2, CONSISTING OF WHITE BLOOD CELLS, PLATELETS AND MALARIAL PARASITE/S (IF ANY) WAS BINARIZED BY DYNAMIC SELECTION OF THRESHOLD VALUE 72

FIGURE 33. FLOWCHART REPRESENTATION OF THE ALGORITHM FOLLOWED FOR MALARIAL PARASITE DETECTION WITHIN RED BLOOD CELLS. 73

FIGURE 34. (ABOVE) THE MALARIAL PARASITE WAS OVERLAID (WITH RED COLOUR) ON THE THIN BLOOD IMAGE (BELOW) THE DETECTED MALARIAL PARASITE WAS PLOTTED ON THE X-Y AXIS FOR CLARITY 74

FIGURE 35. : (A) THIN BLOOD SMEAR IMAGE CONSISTING OF MALARIA PARASITE AT AN ADVANCED STAGE. (B) A MAGNIFIED VIEW OF THE MALARIA PARASITE DETECTED AT AN ADVANCED STAGE. (C) IN COHERENCE WITH THE IMAGE OF THE DIFFERENT WBCS FOUND IN BLOOD , THE IMAGE AS A WHOLE HAS BEEN USED AS AN ILLUSTRATION TO SIGNIFY THAT THE COLOUR PIXELS OF THE MALARIA PARASITE AT AN ADVANCED STAGE ARE MORE CLOSER(IN TERMS OF EUCLIDEAN DISTANCE) TO THE WHITE BLOOD CELLS AS AGAINST THE RED BLOOD CELLS. CLOSE INSPECTION SHOWS THAT, THE MALARIAL PARASITE AT AN ADVANCED STAGE HAS A GRANULATED TEXTURE SIMILAR TO THE TEXTURE OF THE WHITE BLOOD CELL. 75

FIGURE 36. THE SEGMENTATION OF THE NUCLEUS FROM THE WHITE BLOOD CELL 77

FIGURE 37: FLOWCHART OF THE SEMI-SUPERVISED ALGORITHM FOR INFECTION RING IDENTIFICATION WITHIN RBC COMPONENT. (A) A SINGLE RBC CELL WITH P. VIVAX INFECTION IN THE INITIAL STAGE (THE INFECTION IN THE CELL IS HIGHLIGHTED BY WAY OF ROUND MARK), (B) PIXELS AROUND A PARTICULAR INFECTION PIXEL(REPRESENTED BY CO-ORDINATE (2,2) AND MARKED IN RED) HAVING EUCLIDEAN DISTANCE OF $\leq 2\ pixel$ IN TERMS OF PIXEL X-Y CO-ORDINATE IN THE IMAGE. THE MATRIX ALSO PROVIDES THE STRUCTURE OF THE 3 BY 3 WINDOW DRAWN UP WITH A PARTICULAR PIXEL BEING CONSIDERED TO BE AT THE CENTRE OF THE WINDOW(IN THIS CASE PIXEL HAVING CO-ORDINATE 2 BY 2)(C) THE RED(R), GREEN(G) AND BLUE(B) COLOUR DISTANCE VALUE OF EACH PIXEL HAVING AN EUCLIDEAN DISTANCE OF $\leq 2\ pixel$ FROM THE CENTRAL PIXEL COLOUR VALUE(MARKED IN RED). 82

FIGURE 38. DATA MODEL DEVELOPED FOR IDENTIFICATION OF MALARIAL PARASITE WITHIN THIN BLOOD SMEAR IMAGE 86

FIGURE 39. THE ACCURACY VS. NUMBER OF DECISION TREES PLOT TO IDENTIFY THE OPTIMAL NUMBER OF DECISION TREES USED BY THE ADABOOST ALGORITHM 89

FIGURE 40. DATA-MODEL FOR SPECIE AND STAGE CLASSIFICATION 90

FIGURE 41. COMPLETE DATAMODEL FOR THE INITIAL SCREENING PHASE ALGORITHM 95

FIGURE 42. THE ROC CURVE FOR THE ESTIMATION OF THE ALGORITHM PERFORMANCE 101

FIGURE 43. A WBC CLUSTER CONTAINING A WBC AND A VIVAX GAMETOCYTE. (B) COLOUR CODED WBC NUCLEUS AND INFECTION, THE GREEN REPRESENTS THE WBC NUCLEUS. (C) REPRESENTS THE AREA OF THE GAMETOCYTE AND THE WBC,(D) AREA OF THE NUCLEUS THAT HAS BEEN USED AS A FEATURE FOR SEGREGATING WBC FROM MALARIA INFECTION IN WBC CLUSTER 101

FIGURE 44. (A) WBC CLUSTER CONSISTING OF RBC JOINED WITH PLATELET ARTEFACT (B) THE RADIUS FEATURE USED FOR ESTIMATION WHETHER THE RBC SHOULD BE CONSIDERED AS NORMAL OR BIGGER IN SIZE (INFECTION/ OUTLIER) (C) DIFFERENT STAGES OF INFECTION P.VIVAX DETECTED BY THE MAIN PHASE OF THE PROPOSED ALGORITHM 102

FIGURE 45.(A) ORIGINAL IMAGE FROM MAMIC DATABASE, (B) INFECTED RBC CLUSTER WITH RING IDENTIFICATION (BASED ON ALGORITHM REPRESENTED IN FIGURE 37) (C) P. FALCIPARUM INFECTION AT DIFFERENT STAGES AS DETECTED BY THE MAIN PHASE OF THE PROPOSED ALGORITHM 103

FIGURE 46. CONFUSION MATRIX AND PERFORMANCE CALCULATION 108

FIGURE 47. PERFORMANCE METRICS AND CONFUSION MATRIX FOR EACH TYPE AND STAGE OF INFECTION AS OBTAINED FROM THE MAMIC (DATASET #1) 108

FIGURE 48. PERFORMANCE METRICS AND CONFUSION MATRIX FOR EACH TYPE AND STAGE OF INFECTION AS OBTAINED FROM THE ACQUIRED (HOSPITA) (DATASET #2) 109

FIGURE 49. PERFORMANCE METRICS AND CONFUSION MATRIX FOR EACH TYPE AND STAGE OF INFECTION AS OBTAINED FROM THE MIXED DATASET (DATASET #3) 109

FIGURE 50. PERFORMANCE METRICS AND CONFUSION MATRIX FOR EACH TYPE AND STAGE OF INFECTION AS OBTAINED FROM THE MAMIC DATASET FOR WBC CLUSTER (DATASET #1) 110

FIGURE 51. PERFORMANCE METRICS AND CONFUSION MATRIX FOR EACH TYPE AND STAGE OF INFECTION AS OBTAINED FROM THE ACQUIRED DATASET FOR WBC CLUSTER (DATASET #2) 111

FIGURE 52. PERFORMANCE METRICS AND CONFUSION MATRIX FOR EACH TYPE AND STAGE OF INFECTION AS OBTAINED FROM THE MIXED DATASET FOR WBC CLUSTER (DATASET #3) 111

FIGURE 53. PERFORMANCE METRICS AND CONFUSION MATRIX FOR EACH TYPE AND STAGE OF INFECTION AS OBTAINED FROM THE ALL THREE DATASET FOR WBC CLUSTER ... 112

FIGURE 54. CONFUSION MATRIX AND PERFORMANCE OF ADABOOST CLASSIFIER ACROSS THREE DATASET 113

FIGURE 55: REPRESENTATION OF THE ALGORITHMS THAT WERE COMPARED ON THE SAME DATASET BASED ON THE SENSITIVITY & SPECIFICITY VALUES. TK REPRESENTS ALGORITHM PROPOSED BY TEK ET AL [15], DS REPRESENTS THE ALGORITHM PROPOSED BY DAS ET AL.[35], DA REPRESENTS ALGORITHM PROPOSED BY DAS ET AL.[16], RD REPRESENTS ALGORITHM PROPOSED BY ROSADO ET AL.[54], DV REPRESENTS ALGORITHM PROPOSED BY DEVI ET AL.[56] AND NG REPRESENTS ALGORITHM PROPOSED IN THIS RESEARCH WORK 118

FIGURE 56. A DIGITIZED IMAGE OF THE AQUIRED DATASET SHOWING MULTIPLE RBC CLUMPS 120

FIGURE 57. DECLUMPING RESULT OF THE IMAGE IN FIGURE 54. (A) REPRESENTS THE ORIGINAL IMAGE, (B) REPRESENTS THE COLORED CODED IMAGE AND (C) REPRESENTS THE FOREGROUND SEGREGATED IMAGE 121

FIGURE 58. ANOTHER EXAMPLE OF IMAGE WHERE THERE ARE LARGE NUMBER OF RBC CLUMPS AFFECTING THE DECLUMPING PROCESS 122

FIGURE 59. SHOWING A LOW RESOLUTION IMAGE WITH MULTIPLE CLUMPS, (A)THE IMAGE ACQUIRED FROM HOSPITAL. (B) THE MULIPLE COLOURED DECLUMPED IMAGE (C) IMAGE MARKED WITH THE INFECTED REGIONS. 123

FIGURE 60. A HIGHLY STAINED IMAGE FROM THE ACQUIRED DATASETWHERE, (A) IS THE ORIGINAL IMAGE CONTAINING A HIGHLY STAINED CYTOPLASM, (B) IMAGE SHOWING THE FAILURE OF BACKGROUND SEGREGATION AND (C) IMAGE SHOWING RECOLOURED IMAGE THAT HAS OVERLAPPING BACKGROUND AND FOREGROUND 124

LIST OF TABLES

TABLE 1. TABLE SHOWING A COMPARATIVE ANALYSIS OF MALARIA STATISTICS FOR YEARS 2016-2018 DOCUMENTED AND DISTRIBUTED ANNUALLY BY WORLD HEALTH ORGANIZATION 5

TABLE 2 COMPARISON OF MORPHOLOGICAL CHARACTERISTICS OF PLASMODIUM SPECIES DURING DIFFERENT STAGES OF ITS LIFECYCLE 7

TABLE 3. COMPARISON OF PBS TECHNIQUE WITH RDT 14

TABLE 4. THE PARAMETERS FOLLOWED FOR DATASET DEVELOPMENT WITH THE DESCRIPTION OF DIFFERENT PARASITES AS OBSERVED BY EXPERTS (GROUND TRUTH) 49

TABLE 5. THE EXHAUSTIVE LIST OF THE FEATURES THAT WERE CALCULATED FOR EACH OF CLOSED COMPONENTS PRESENT WITHIN CLUSTER 2 77

TABLE 6. LIST OF SELECTED FEATURES FOR CLASSIFICATION 78

TABLE 7. FEATURE LIST USED FOR SEGREGATION OF WBC CLUSTER FROM OTHER CONNECTED COMPONENTS 80

TABLE 8. FEATURE LIST USED FOR SPECIE AND STAGE CLASSIFICATION FROM RBC CLUSTER 84

TABLE 9. THE ACCURACY, SENSITIVITY AND SPECIFICITY VALUES WHEN THE NEIGHBOURS USED FOR CLASSIFICATION OF A TEST DATAPOINT WERE VARIED BETWEEN 1, 3 AND 5 93

TABLE 10. COMPARISON OF PERFORMANCE METRICS BETWEEN RULE BASE AND ML BASED ALGORITHMS 94

TABLE 11. COMPARATIVE STUDY OF THE OVERALL PERFORMANCE OF THE PROPOSED SCREENING ALGORITHM WITH OTHER RESEARCH WORKS 96

TABLE 12. COMPARATIVE STUDY OF THE OVERALL PERFORMANCE OF THE PROPOSED SCREENING ALGORITHM WITH OTHER PROPOSED METHODS BY AUTHORS USING 250 IMAGES OF DATASET USED IN THE THESIS. 97

TABLE 13. COMPARATIVE ACCOUNT OF THE ACCURACY ACHIEVED BY THE TWO METHODS USED FOR IMAGE BACKGROUND SEPARATION BEFORE AND AFTER ILLUMINATION CORRECTION (WHICH WAS SEPARATELY IMPLEMENTED FOR THE TWO DATASETS AT HAND) 98

TABLE 14. PERFORMANCE EVALUATION FOR CLUMP IDENTIFICATION USING TWO SEPARATE THRESHOLDS 99

TABLE 15. RED BLOOD CELL SEGREGATION FROM WHITE BLOOD CELL IN THE DIGITIZED THIN BLOOD SMEAR IMAGE DATASET – PERFORMANCE METRIC USING TWO DIFFERENT THRESHOLD (FOR 1ST ITERATION). 99

TABLE 16. RED BLOOD CELL SEGREGATION FROM WHITE BLOOD CELL IN THE DIGITIZED THIN BLOOD SMEAR FOR THE TWO DATASETS UNDER CONSIDERATION 100

TABLE 17. DETECTION ALGORITHM PERFORMANCE EVALUATION METRIC 100

TABLE 18. THE VALUES OF THE GLCM FEATURES THAT ARE SIGNIFICANT TOWARDS DISTINCTION BETWEEN VIVAX SCHIZONT VIVAX GAMETOCYTE FROM MAMIC DATASET 106

TABLE 19. PERFORMANCE STATISTICS FOR IDENTIFICATION OF INFECTED RBC FROM NORMAL RBC CELLS AND FOR IDENTIFICATION OF INFECTED WBC CELL CLUSTER USING ONE VS ALL STRATEGY 106

TABLE 20. PERFORMANCE STATISTICS FOR FINAL AVERAGE CLASSIFICATION ACCURACY ACROSS ALL INFECTION CLASSES WITH SINGLE AND ENSEMBLE CLASSIFIERS (AT IMAGE LEVEL) 107

TABLE 21. COMPARATIVE STUDY OF THE OVERALL PERFORMANCE OF THE PROPOSED DETECTION AND CLASSIFICATION SYSTEM WITH OTHER COMPARABLE METHODS PROPOSED BY DIFFERENT AUTHORS. (ONLY JOURNAL PAPERS ARE CONSIDERED FOR COMPARISONS) 117

TABLE 22. COMPARATIVE STUDY OF THE EXECUTION TIME OF THE PROPOSED DETECTION AND CLASSIFICATION SYSTEM WITH OTHER COMPARABLE METHODS PROPOSED BY DIFFERENT AUTHORS. (ONLY JOURNAL PAPERS ARE CONSIDERED FOR COMPARISONS). N: NUMBER OF DATAPOINTS, D: NUMBER OF FEATURES (FEATURE DIMENSION), K: NUMBER OF K VALUE FOR NEAREST NEIGHBOUR/CLUSTER AND C: NUMBER OF CLASSES... 117

TABLE 23. TABLE SHOWING COMPARISON OF COST BENEFIT ANALYSIS OF THE EXISTING MANUAL METHOD AND AUTOMATED CAD SYSTEM (PROPOSED) 127

LIST OF ACRONYMS USED

ACRONYM	FULL FORM
ACC	Automated Cell Counter
ANN	Artificial Neural Network
ART	Adaptive Resonance Theory
CAD	Computer Aided Dignosis
CDC	Centre for Disease Control and Prevention
CT-Scans	Computed Tomography Scanners
DICOM	Digital Imaging and Communications in Medicine
DLL	Depolarized Laser Light
DNA	Deoxyribonucleic Acid
FDA	Food and Drug Administration
FF-BPNN	Feed-forward Back Propagation Neural Network
GIS	Geographical Information System
GLCM	Grey Level Co-occurrence Matrices
HIS	Hue Intensity Saturation
HSV	Hue Saturation Value
IFA	Immunofluorescence Antibody Testing
k-NN	k-Nearest Neighbor
LAMP	Loop Mediated Isothermal Amplification
LDMS	Laser Desorption Mass Spectrometry
LED	Light Emitting Diode
LoG	Laplacian of Gaussian
MCS	Multiple Classifier System
MKM	Moving K-Means
MLP	Multi-Layered Perceptron
MQBC	Modified QBC
MRI	Magnetic Resonance Imaging
MS	Mass Spectrometry
PACS	Picture Archiving and Communication system
PBS	Peripheral Blood Smear
PCA	Principal Component Analysis
PCR	Polymerase Chain Reaction
PNN	Probabilistic Neural Network
QBC	Quantitative Buffy Coat
QFT	Quaternion Fourier Transform
RBC	Red Blood Cell
RBF	Radial Basis Function
RDT	Rapid Diagnostic Test
RGB	Red Green Blue
ROC	Receiver Operating Characteristics
ROI	Region of Interest
SUSAN	Smallest Univalve Segment Assimilating Nucleus
SVM	Support Vector Machine
US	United States
WBC	White Blood Cell
WHO	World Health Organization
WSI	Whole Slide Imaging Scanners
YC_BC_R	Luma Blue Difference Chroma Red Difference Chroma

CHAPTER 1
INTRODUCTION

1.1. Overview

Image processing and image analysis is a core area of application in computer science. Several research work has contributed and consolidated this domain of research. With the progression and advancement of digital imaging, contribution towards meaningful assessment of images have gained importance. An image captured by some imaging modalities needs to be understood and deciphered to gain insight on the image/s and hence extract pertinent information from the captured image/s. Images can be captured using digital camera, scanners, telescopes, microscopes and other electro-optical devices used in different industries and domains. Similarly images are also generated by medical imaging hardware like Computed Tomography Scanners (CT-Scans), Magnetic Resonance Imaging (MRI) machines, X-ray machines and digital Whole Slide Imaging Scanners (WSI).

The images obtained require processing so that meaningful information can be retrieved and collated. Computerized algorithms play a significant role to achieve this goal. Manual interpretation of image may vary based on the perception of details and reasoning capability of the operator. Human bias and fatigue are two key elements that impede the performance of manual recognition and interpretation of images. However, the advent of computerized image processing tools has helped to overcome these impediments by assisting the operators in image analysis.

Computerized Image processing or Digital image processing is often referred to as Computer Vision and found its initial application in the study of satellite imagery captured by spy satellites to track enemy movement. Though initial applications of Computer Vision and Pattern Analysis was in defense related research, but now this has found applications in every walks of life. Digital image processing have wide range of application ranging from industrial manufacturing, Geographical Information System (GIS) and also in Medical imaging. Digital image processing and automated analysis of images have significantly contributed to the betterment of several processes that affect human lives.

This research work particularly contributes towards medical image processing. Medical images are generated by several imaging modalities. Such images are interpreted and analyzed by some medical professional. Based on the source and imaging modalities the professional may be a radiologist (in case of X-rays, CT, MRI imaging modalities) or a pathologist (microscopic images pertaining to a cell or tissue sample). However, a large number of images are generated that is needed to be analysed with limited number of human resource. Automated computer application or Computer Aided Dignosis (CAD) systems can provide adequate analysis that can multiply and maximize the limited human resource for correct analysis. Effective, timely and accurate analysis contribute to correct inference that is vital for diagnosis of disease.

This thesis contributes to the knowledge domain of medical image processing. The contribution is aimed at developing automated CAD applications for Pathology. Malaria disease identification is an age-old problem that significantly affect a huge population. The research work presents an approach for automated analysis of Malaria infected slides. The subsequent sections of this introductory chapter describes the background study and information related to various aspects pertaining to Malaria. The final section of this chapter provides an insight of the overall organisation of the thesis.

1.2. Digital Pathology

Pathology is a branch of medicine that combines the science of disease, their cause, effect and diagnosis. A Pathologist determines the cause of a particular disease conditions based on certain prescribed tests (chemical/clinical/microscopy), for accurate diagnosis and provide relief to the suffering patient. Most of these test are conducted by automated equipment using body fluids/tissue sample extracted from the patient. Microscopy plays a vital role in disease determination for cases of parasitic invasion within tissues and to locate abnormality in histological/cytological body samples.

Microscopic examination of cellular and histological samples are widely used as a basis for disease detection. However, with the introduction of advanced digital microscope and high resolution scanners the approach towards pathology had a paradigm shift towards 'virtual microscopy' as an innovation in diagnostic workflow. Handling of glass slides across the labs is cumbersome and susceptible to loss of the slide or decreased quality of the specimen. The associated turnaround time (from the sample collection to report generation) is time consuming. With the increase in reliability of digital equipment like digital imaging technologies, computer hardware and software, there has been persistent acceptance of Digital Pathology in the medical community [1].

Conversion of a biological specimen in a glass slide to an image is referred to as Virtual Microscopy. This is the first step towards digital pathology. High resolution Whole Slide Imaging Scanners (WSI) perform this task at resolutions of 40-60X magnification. A high resolution image obtained of the whole slide can be extended to the size of a tennis court when projected/printed at 300 dpi resolution [2]. The information can be archived for training and technical education purpose, telemedicine applications, primary/secondary diagnosis or for a second opinion, review of consultation and for quality assurance mechanisms [3]. The consulting pathologist will view the slide images on high resolution monitors that are specialized for medical purpose. The pathologist will be able to analyse remotely at the time of his preference without the sample being affected/destroyed or stained sample getting discolored. The images can be distributed among consultants for double review or expert review and can result in faster workflow in pathological laboratories.

Digital pathology using WSI has been granted certification by Food and Drug Administration (FDA) [4] in 2017 for its application in primary diagnosis of disease. Medical images obtained from different equipment/vendors and of different modalities have been standardized by the Digital Imaging and Communications in Medicine (DICOM) standard. The DICOM images and Picture Archiving and Communication system or PACS system are being evolved to accommodate WSI for medical diagnosis. A working group, WG-26, was established by DICOM for this very purpose in 2005 [5]. Several studies indicate that the performance of WSI and glass slide is similar [6].

The use of digital imaging has also opened a new technological dimension for pathology. The images can be processed with Artificial intelligence and machine learning algorithms in Computer Aided Diagnosis Systems or CAD system. Such a system can be used to identify abnormalities independent of human intervention and is referred as the third revolution in pathology [7].

1.3. Malaria Disease

Malaria is the oldest and cumulatively the deadliest of the human infectious diseases and is a primary cause of child mortality. The disease is predominantly widespread in tropical climatic regions that are backward and under-developed. Female Anopheles mosquito is the sole vector for the protozoan infectious disease that affects human population. The disease attains epidemic proportions in remote rural areas within a very short time. Prevention of the disease by curtailing the breeding grounds of the vector has proved futile in most areas of the world. The disease can only be managed with early detection, confirmation of species type, stage and density of parasite within the human blood.

Malaria is often designated as the 'King of Diseases' [8] [9] due to its predominance in the world of infectious diseases causing human mortality. The disease was a cause of dread for ages and still continues to be a threat to humankind for being the fifth deadliest infectious disease [8]. The disease was known to the medical practitioners for more than 50,000 years [10], however its effect on humans were identified at a later stage. The disease was known to almost all of the ancient civilizations of the world including Chinese, Indian, Greek, Babylonians and Romans. Various historical documents, scriptures and medical text finds the mention of a disease that is seasonal, causes intermittent fever, originates from marshy unhygienic areas, vector borne and is responsible for mortalities of epidemic nature [9]. The disease finds its ways in the medical writings of Hippocrates, the Greek physician and also in the Vedic texts, 'Shusruta Samhita', the Canon of Medicine and the ancient Chinese 'Nei Ching' [9]. The origin of the disease is prehistoric and its transmission to humans happened millions of years ago [11]. In the book of Robert Sallares [12], the origin of the name for the disease and its impact on the ancient Italian society was discussed. The Etymology of Malaria originated from Italian 'mal aere', that has the meaning of 'bad air' found in the book entitled 'Scitture della laguna' of Marco Cornaro [13], published in 1440 in the city of Venice. However, the term Malaria got its introduction in English literature from the letters of Horace Walpole to his cousin [13]. The word got associated to the specific disease in the publication of a book by Guido Baccelli called 'La Malaria di Roma' in 1878 [13].

1.3.1. Socio-economic Impact

The disease have profound impact on the livelihood of people in endemic regions of the disease. The extensive economic effect of the disease is not only felt on the household but they tend to affect the financial infrastructure of nations and are definitely responsible for the economic backwardness of the countries. This can further be attributed to the fact that insufficient resources, lack of basic amenities, limited skills, insecurity and lack of power can be related to poverty. This leads to poor access to health care in such places [14]. Poor living conditions like overcrowding and improper housing increase susceptibility towards various infections including Malaria [15]. Malnutrition among women and children results in the poor physical development of children who become prone to disease like Malaria. Illiteracy is profoundly present among poor which prevents them from being aware of health related issues. Lack of sufficient healthcare infrastructure available in remote areas without adequate equipment, medicine or trained staff [16] resulting in low-quality health service often becomes insufficient to tackle Malaria outbreak. Malaria causes financial loss incurred by the family due to direct costs (cost of treatment and medication), indirect costs (loss of working hand) and opportunity costs (the financial gain by an earning individual). The citation [17] reports that an estimated Mean Direct Cost on Malaria treatment by a household can be up to 2% - 2.9% of annual income The disease is responsible for making poor countries even poorer with the decrease in

Gross Domestic Product (GDP) value. Studies by [18] suggest that morbidity due to Malaria infection reduce the annual per capita growth by 0.25% points in Malaria endemic countries. Economic review by Sachs and Malaney concludes that "where Malaria prospers most, human societies have prospered least" [19].

1.3.2. Malaria Statistics

The World Health Organization (WHO) under the aegis of United Nations collects Malaria related data from all over the world. They are the pioneers of Malaria prevention and eradication programs in both poor and underdeveloped nations in the world that are worst affected by the disease. Malaria related data is obtained from all member nations. The data is obtained by WHO, working closely with government hospitals and care centers from urban to rural and even from remote and inaccessible locations. The data is compiled and projected on the official website. The reports released each year reflect the efforts made by the member states towards Malaria eradication, prevention and awareness. The Table 1 summarizes the statistical details for last three years.

	Report 2018 [20]	**Report 2017** [21]	**Report 2016** [22]
Data Representation (Year)	2017	2016	2015
No. of cases Globally in Millions	219 (95% CI)	216 (95% CI)	212
Highest No. of Cases	African Region 92%	African Region 90%	African Region 90%
Percentage of case in India (Rank)	4% (4th)	7%< (NA)	7%< (NA)
Mortalities Globally	435000	445000	429000
Percentage of fatalities (5<)	61%	Not Reported	70%
Percentage Mortalities in India	4%	6%<	6%<
Complete Elimination	Paraguay	Kyrgyzstan, Sri Lanka	10 Countries reported no fatalities

Table 1. Table showing a comparative analysis of Malaria Statistics for Years 2016-2018 documented and distributed annually by World Health Organization

1.3.3. Biological Aspect of Malaria Disease

The Malaria parasites belong to the Phylum Apicomplexa, Class Sporozoea, Sub-Class Coccidia, Order Eucoccida and Sub-Order Haemosporina. Under this sub-order is Genus *Plasmodium*. This genus is characterized by the presence of two hosts in the lifecycle with Schizogony (asexual cycle) and Sporogony (sexual cycle). There are 5 species that cause Malaria infection to human beings. The disease is most prevalent in the tropical regions of the world between 60° N and 40° S. The parasite resides in two hosts. Within the human host the parasite undergoes asexual lifecycle and within female Anopheles mosquito, it undergoes sexual lifecycle.

The female Anopheles mosquito is key to the widespread dispersion of the Malaria disease. Infected mosquitoes that had sucked the blood of an infected patient initiates and becomes the carrier of the disease without itself being affected by the disease. The parasites travels to the mosquito gut and multiplies its numbers to create a pool of parasites that migrates to the saliva of the vector for further transmission to healthy human host.. The blood sucking event of such infected mosquito inoculates the parasite in the human vascular system. The *Plasmodium* parasite finds the new host to propagate its progeny. The human host often becomes a reservoir for the parasite where they live multiple generations and increase their numbers at an exponential rate at the cost of human red blood cells. The parasites after entering the human blood stream, travel to the liver. The hepatic tissue is the first region of the body that gets infected. The symptoms of Malaria is not observed at this point of time. During this gestational period the parasite multiplies itself to infect the red blood cells and also to defeat the host immune system. The host especially children with malnutrition are susceptible due to poor immunity. After invading the red blood cells the parasites starts its asexual lifecycle and between every 48 to 72 hours, the parasites mature and destroys the host red blood cell. The cell bursts to release a colony of newly formed parasites that is ready to carry on further invasion of new, healthy host red blood cells. The parasite exhibits polymorphic life forms within the human host indicating different lifecycle stages within the host. The bursting of cells typically associated with severe shivering and cramps exhibited by the patient. Typical Malaria symptoms of intermittent high fever is manifested by the host that occur in cycles, repeating the process every two to three days. The parasite multiplies at the cost of human red blood cells resulting in blood loss and Anemia related issues. Progression of the disease, if not controlled, leads to severe blood loss conditions, organ failure and mortality. The most virulent specie, *Plasmodium falciparum* among the five species of parasites known to infect humans; infection spreads to fine blood capillaries of brain, where they block the flow of blood, resulting in rapid loss of brain functioning and mortality. The other species causes severe illness and suffering but mortality is less reported. This specie, *P. falciparum* is mostly reported in sub-Saharan Africa and Oceania. However, latest reports indicate that in India this species has progressed to be the leading cause of Malaria infection relegating *Plasmodium vivax* to the second place. The Centre for Disease Control and Prevention (CDC) of United States (US) government [23], create awareness for this disease along with other infectious disease.

INTRODUCTION

Malaria protozoa exhibit several forms within its lifecycle. The parasite infects the human system as Sporozoite. Among the five species known to infect human kind, P. knowelsi is little known to affect and cause disease. The two species namely *P. vivax* and *P. falciparum* constitutes the primary form of infection in the Indian sub-continent and hence considered for this research work. The Table 2, compares the differences between the polymorphic life forms of the parasite within a human host. Figure 1. shows the life-forms in different stages recorded by a digital microcope. These images are obtained from CDC [24] and these images are distributed freely for academic purpose.

Plasmodium	*P. vivax*	*P. falciparum*
Common Name	**Benign Tertian Malaria**	**Malignant Tertian Malaria**
Trophozoite	Early Trophozoite have blue cytoplasmic ring, red nuclear mass & vacuole. RBC enlarges and irregular with Schüffner dots.	Early ring form with fine and uniform cytoplasm ring with nucleus lying outside the ring often divided into two parts and at opposite pole. RBC remains normal with 6-12 Maurer's cleft.
Schizont	They almost fill the enlarged RBC, the nucleus is large and lies on the periphery. After nuclear division on average 16 daughter individual form a rossete like cluster. RBC burst at this stage	They fill two third of RBC. After nuclear division 8-32 daughter cell produced. RBC remains un-enlarged. They burst to release the cells.
Merozoite	12 - 24 oval cytoplasm containing mass.	18 - 24 circular cytoplasm mass
Gametocyte	Spherical in shape and slightly enlarged RBC containing granules	Host RBC is filled and the gametocyte is Crescent shaped

Table 2 Comparison of Morphological characteristics of Plasmodium species during different stages of its lifecycle.

INTRODUCTION

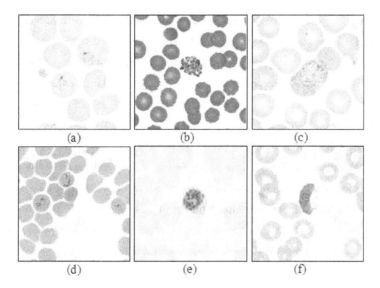

Figure 1. A comparison of the different lifecycle forms of Plasmodium genus. Figure (a)-(c) are Trophozoite of P. vivax, Mature Schizont of P. vivax, Gametocyte of P. vivax and (d)- (f) Trophozoite of P. falciparum, Mature Schizont of P. falciparum, Gametocyte of P. falciparum. The images obtained from CDC, DPDx- Malaria Image Library [24]

1.3.3.1. Plasmodium Lifecycle

The Malaria parasite inhabits two different hosts in its lifecycle characteristic to the genus. The asexual phase within the human is initiated by the injection of sporozoites through a blood sucking event by an infected mosquito. Within the human hosts, it undergoes three to four distinct cycles depending on the infected species. A Pre-Erythrocytic or Primary Exo-Erythrocytic Schizogony happens within the parenchymal cells of hepatic tissues of the liver. A single generation multiplication happens and merozoites are liberated called cryptozoites. The smaller micromerozoites enter the blood circulation to initiate the next cycle. Within the blood, the merozoites infect the RBC and the manifestation of the disease initiates after a definite number of days specific to each species. The Erythrocytic Schizogony cycle of development of Trophozoites, Schizonts, Merozoites and re-invasion of fresh RBC continues for several cycles at definite interval depending on the species. This is characterized by the recurrent febrile symptoms of the patient. This continues till the exhaustion of asexual life of the parasite. After achieving a certain number of asexual reproduction, the merozoites develop into gametocyte or the initiation of the sexual phase or gametogony. They generally develop within the smaller blood vessels of the internal organs like spleen and bone marrow. Mature gametes are found on the peripheral blood vessels ready for transmission by disease vector. A

INTRODUCTION

Latent Hepatic stage is observed in species like *P. vivax* and *P. ovale* where the merozoite enter a suspended state called Hypnozoite responsible for relapse of the disease.

Soon after a blood sucking event by female Anopheles mosquito from a Malaria infected person the sexual cycle or Sporogony is initiated. A single blood meal requires more than 12 gametes/mm^3 of which the female macrogametocytes should exceed the male microgametocytes. Fertilization occurs to form a zygote which matures to form an Ookinete. This process occurs within the mid-gut of mosquito. The ookinete further matures and form oocyte on the stomach wall. Within the oocyte several meiotic and mitotic division results in the development of thousands of sporozoites. The oocyst raptures and the released sporozoites travel all around the mosquito body through the body fluids. They have an affinity to be present within the salivary glands for easy transmission to the intermediary host via a bite on human tissue [25]. Figure 2, shows the schematic diagram of the parasite lifecycle.

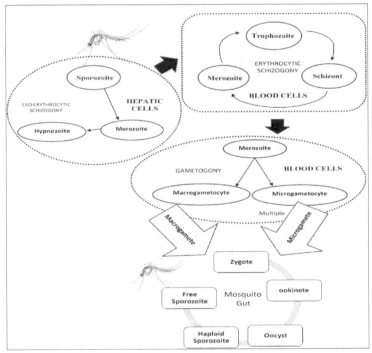

Figure 2. The life cycle of Plasmodium genus and its development in Human and Mosquito hosts.

1.3.3.2. Composition of Blood

In mammals, oxidation of tissues is performed by binding the oxygen molecules to the haemoglobin contained within the circulating discoid shaped and non-nucleated Red Blood

Cells (RBC) or Erythrocytes. Malaria infects and lives a parasitic life within the host RBC and destroys the same eventually. For computerized algorithms identification and differentiation of infected and normal RBC is vital. The possible hinderence in the segmentation process are the presence of White Blood Cells (WBC) or leukocytes that are the primary defence system of the body against infections. In circulation of blood, almost 7000 white blood cells per microliter are present in normal human being. A healthy adult individual will normally have 1% of WBC in the total blood by volume as its constituent. The WBC is further classified on the basis of the presence/absence of granules (these are digestive enzymes or other chemical substances contained in sac like structures) within the cytoplasm. Each type of WBC is classified as to be a granulocyte or a-granulocyte. Neutrophils, Basophils and Eosinophils are Granulocytes that differ in the lobular structure of nucleus. Lymphocytes and Monocytes have single lobed nucleus and void of granules and constitutes Agranulocytes. Figure 3 shows all the different types of WBC that are found within a smear slide image.

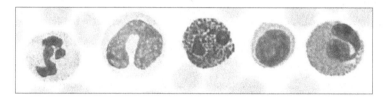

Figure 3. The figure shows five different types of WBC with Neutrophil at the leftmost position followed by Monocyte, Basophil, Lymphocyte and Eosinophil at the extreme right. This rare image was recorded by a Haematologist while scanning through smear slides.

1.3.4. Diagnosis of Malaria

Early and effective treatment of Malaria can prevent its rapid spread among the immediate geographic location. The diagnosis of Malaria identifies the presence of Malaria parasite cells, antigens and antibodies within the human blood. There are different Malaria lifecycle stages and five types to identify from. Early diagnosis can prevent mortality rates. Diagnosis of Malaria depends on the availability of appropriate diagnostic equipment and presence of adequate trained technicians.

1.3.4.1. Clinical Diagnosis of Malaria

The clinical diagnosis of Malaria is done by a medical practitioner, very often at low cost or sometimes even free of cost in government hospitals and medical centers. Malaria is often characterized by non-specific symptoms at an initial stage of infection like fever, headache, weakness, chills, dizziness, abdominal pain, diarrhoea, nausea, vomiting, anorexia, and pruritus [26]. Such symptoms are similar to a wide range of bacterial or viral disease. To distinguish

Malaria from other diseases prevalent in the region more specific diagnostic methods are required. The citation by Kyabayinze et al. [27] also confirms that improvement of accurate Malaria diagnosis is possible by combining clinical and parasite-based detections.

1.3.4.2. Laboratory Diagnosis of Malaria

The laboratory diagnosis of Malaria is the most widely used procedure for Malaria detection in the region where Malaria commonly occurs. The laboratory techniques vary in the nature of tests conducted but generally, they involve the collection of blood sample from the affected patient and make use of scientific equipment or chemicals to identify the presence of Malaria parasite. Such laboratory tests should be done carefully and with experienced technicians. The symptoms of Malaria are often non-specific in nature so wrong diagnosis in Malaria prone regions resulting in over/under treatment may occur as well as a false negative diagnosis in the non-Malaria region is a possibility [28].

1.3.4.2.1. Microscopic Diagnosis Using Stained Thin and Thick Peripheral Blood Smears (PBS)

Since the discovery of Malaria by Laveran in 1881, use of conventional light microscopy was the primary method on blood smear slide stained with Giemsa, Wright's, or Field's stains [29]. Staining method was discovered by Romanowski in 1891 for enhancing the parasite for better identification. The combination of Giemsa-stained thick smear slide for initial screening and thin smear slide for species identification still remains the 'gold standard' for laboratory diagnosis of Malaria [30]. Both type of smear, thin and thick can be used to calculate Parasitaemia [31]. If the test reports negative detection the process is repeated every 8 hours for a couple of days to confirm absence of the disease [32]. The slide preparation is vital to the method. It requires the involvement of skilled technicians (phlebologist) and the process starts with a prick on the fingertip (sterilized with ethyl alcohol). Only two drops of the oozing out blood is taken on an oil-free glass slide to prepare the smear. To obtain a thick smear on a slide the blood droplets are dabbed with the corner of another slide to form a circular region. The same is allowed to air-dry and Giemsa staining of 1:20 volume is applied for 20 mins. The excess stain is washed and the slide is placed vertically to dry out. The thin smear requires another slide to be held at 45 degrees and swiftly moved across along the length of the slide. The slide is dried and fixed with methanol and stained with Giemsa staining of 1:20 volume for 20 mins. The same is washed and kept for drying. The slides prepared are then examined under a light microscope by pathologists [33]. The WHO documents the procedures to be followed by a laboratory technicians while detecting Malaria in the practical microscopy guide [31].

INTRODUCTION

The advantage of this system can be attributed to its simplicity and economically viability. The visual differentiation between normal/ parasite infected RBC, the species causing the disease and the Parasitaemia calculation helps in effective disease control. The preparation of slides is however laborious, time taking and requires the laboratory technicians to be skillful. The challenge still lies on the pathologists to identify parasite even at the low level of infection. The primary disadvantage of such system is low sensitivity at detection level and species identification at low Parasitaemia. An expert technician though can detect 5 parasites/µl, on an average 50-100 parasite/µl can be detected by technicians [34]. The absence of skilled technicians in areas where Malaria is less frequently reported, availability of experts in remote areas and in cases of asymptomatic Malaria or low levels of Parasitaemia contributes towards low reporting rates [35]. Microscopy may not be available in remote areas where there is no medical infrastructure or basic health care and/or absence of electricity. They are also useless when considered for high throughput requirements [36]. Similarly; negative detection may arise where blood is drawn from patients on anti-Malaria drugs, during the apyrexial interval of infection and the first couple of days of primary infections. The presence of artefact makes the visual detection problematic. Anything other than blood component or a parasite is considered as an artefact. They may include bacteria, spores, stain crystals and dirt [31]. Other elements that may be found within blood are other blood parasites, RBC anomalies like Howell-Jolly bodies and reticulocytes are often reported as artefacts.

1.3.4.2.2. Quantitative Buffy Coat Technique (QBC)

The quantitative buffy coat test was devised to simplify and enhance detection of Malaria parasite [37]. This method involves the collection of blood in haematocrit capillary tube coated with anticoagulants and Acridine orange fluorescent dye. The tube is centrifuged for 5 mins at 12000 G. The dye stains the DNA of the parasite for observation through fluorescent microscopy [33]. Ultraviolet light is used to observe the region between red blood cells layer and buffy coat. Efficacy of the QBC over thick PBS achieve greater than 90% sensitivity for species identification. The parasite DNA is observed as bright green. QBC is found to be less sensitive to non-falciparum species [38]. Modern portable fluorescent microscope with Light Emitting Diode (LED) along with slides coated with dye has made QBC popular. The research citation [28] on Modified QBF Method (MCBF) where the buffy coat is obtained by micro-centrifuge at 6000 rpm. The blood capillary tube is cut using Adam's plier. A steel wire was used to remove and push the sediment column at the junction of plasma and buffy coat of blood to a glass slide. A thin and thick smear was prepared and compared with normal thin and thick smear slides. Slides obtained from 100 suspected Malaria patients were analysed and compared. It was concluded that normal smear slide had 86.79% sensitivity and 100% specificity whereas with (MCBS) improved its sensitivity from 86.79% to near 100%, the process can easily be performed and is affordable. QBC method is simple, dependable and user-friendly. However, the equipment required is costly. They cannot calculate the Parasitaemia or determine the species. Moreover, the detection rate decreases with non-falciparum species [39].

1.3.4.2.3. Rapid Diagnostic Tests (RDTs)

The Rapid Diagnostic Test (RDT) kit detects Malaria quickly. WHO held a congress and prepared a document, being reference number WHO/MAL/2000.1091, entitled "New Perspectives in Malaria Diagnosis". The report emphasized that results of RDT should be similar with conventional PBS method by normal technician under field conditions. Similarly, the results obtained should have sensitivity above 95% compared to microscopy findings with a Parasitaemia level of 100 parasites/l (0.002% Parasitaemia). Malaria parasites produce specific antigens proteins like Histidine-Rich Protein II (HRP-II) or Lactate Dehydrogenase (LDH) that circulates in the blood stream of the infected patient The RDT's chemicals mark the presence of such antigens in the blood sample provided. They can detect a single species or multiple species. The results are shown immediately. There is more than 200 Malaria RDT kits on the market based on similar principles but they test with different chemicals and for different antigens. Currently, 86 Malaria RDTs are available from 28 different manufacturers [40]. Manufacturing of RDT kits have significantly expanded around the world. The RDT developed initially for detection of falciparum has now been adapted for vivax infection [41],[42] and [43]. A new RDT product launched to detect knowlesi species [44].

A survey undertaken by WHO and reported in World Malaria Report 2014 [45], that a total of 319 million RDT was sold by manufacturers in 2013 where 59% of P. falciparum specific tests and 39% for multiple species test. RDT has established itself as an alternative approach to microscopy and have shown high sensitivity and specificity. For RDT the Mean Operational Sensitivity with reference to microscopy was 64.8%. Higher Sensitivity of RDTs was proportional to increasing in Parasitaemia. Even with poor slide quality the Specificity of 87.8% was achieved [46].

SD Bioline Malaria Antigen test [41] used for *vivax* Malaria detection. Data obtained from 732 patients living in areas where Malaria is endemic. The performance statistics showed a sensitivity of 96.4%, and its specificity was 98.9%. Among the sample 95.4% patients with Malaria showed a 100% positive predictive value for *vivax* Malaria. The research citation [47] performed a comparative analysis of SD Bioline Malaria Ag Pf. According to authors there are

86 Malaria RDT products from 28 different manufacturers. They are all based on the same principle and use antibodies that detect only three groups of antigens. According to research citation [48] a popular RDT brand OptiMAL dipsticks detection effectiveness was estimated to have sensitivity of 96.6% and specificity of 85.4%. The positive predictive value of 92.7% and the negative predictive value of 92.6% was obtained by performance analysis.

The use of RDT is more prevalent for regions highly prone to Malaria but lacks basic medical infrastructure, electricity etc. WHO is contemplating to develop quality control guidelines to boost the confidence of people for the use of these products [40]. The simplicity and reliability of RDT have been enhanced so that it can be used in rural areas and regions where Malaria is generally unknown or imported [49]. Though RDT do not require specialized training or equipment the detection sensitivity is less than microscopy, nor they can distinguish among species and cannot calculate Parasitaemia [30]. The comparison of RDT with PBS technique is summarized in Table 3.

Parameters	Peripheral Blood Smear and Light Microscopy (Gold Standard)	Rapid Diagnostic Kit (RDT)
Minimum Parasitaemia Required	50 parasite/µl	> 100 parasite/µl
Parasitaemia Determination	Yes	No
Stage Differentiation	Yes	No
Species Identification	Yes	No
Ability to classify species	Yes	Partially by certain products
Average Time of Detection	30 min	5 min
Cost of Detection	Low for bulk screening	Medium
Trained Technician	Yes	No
Infrastructural Requirements	Yes	No
Special Laboratory Reagents	Yes	No
Variability of Detection	Inconsistent	Mostly consistent

Table 3. Comparison of PBS technique with RDT

1.3.4.2.4. Serological tests

Serological tests refer to detection of specific antigens and antibody from chemical assay of blood plasma. Immunofluorescence Antibody Testing (IFA) for Malaria detection has been proved relevant in contemporary findings [50]. The citation [51] demonstrated consistency of IFA so that it can be considered as 'gold standard' in serology testing for Malaria. It was observed that post infection of two weeks the antibody lasted for 3-6 months. The serum prepared on slide was titrated and observed under fluorescent microscope. Titers greater than

1:20 are considered positive whereas less than 1:20 was considered unconfirmed [33]. Currently, in developed countries IFA is used for screening in blood banks to reduce transfusion related Malaria [52] and [53]. The primary advantage of using this method is that it is extremely sensitive and specific [54]. But the process is manual with only limited number can be processed by trained experts and viewed by costly fluorescence microscopy. Similarly, the system is yet to be standardized so that it can be correlated with findings of other laboratory.

1.3.4.3. Molecular Diagnosis of Malaria

The recent advancement in molecular biology and biotechnology has contributed to the development of modern techniques and machines that are used for different types of pathological testing and determination of diseased conditions. These modern techniques involve costly machines that can be adapted to detect Malaria infection. Some of these techniques adapted for Malaria identification are described in the following subsections.

1.3.4.3.1. PCR Technique

Polymerase Chain Reaction (PCR) method can amplify trace amounts of DNA present in any body fluids for diagnostic purpose. Traces of elements containing segments of DNA or RNA can be multiplied by copying the strands thus increasing the probability of detection of the small amount of pathogenic DNA. Such techniques have helped in molecular diagnosis of Malaria particularly in cases where they are admixed with other pathogens and cases of very low Parasitaemia levels [55]. This technique is growing in its popularity for definite confirmation of infection, disease management and to detect specific resistance to medicine [33]. PCR has greater reliability and accuracy in determination of infection with a comparison to RDT and QBC [56]. PCR technique performs very well in cases of Parasitaemia levels of $\leq$ 0.0001%, i.e. 10-20 times lower than conventional microscope [57]. The advantages of PCR technique are in achieving best possible sensitivity and specificity values, its ability to detect at low Parasitaemia levels and detect drug resistance. The disadvantage of PCR is however that it requires complex equipment that is costly and requires trained experts. This cannot be implemented in rural Malaria endemic areas [58] and cannot be implemented in developing nations due to its inherent complexity and costly maintenance [59].

1.3.4.3.2. LAMP Technique

The Loop Mediated Isothermal Amplification (LAMP) is very similar to PCR and performs gene amplification. It has been widely used for early detection of microbial diseases. This is a relatively low cost method for Malaria diagnosis. The LAMP technique detects the presence of *falciparum* parasite by locating the occurrence of 18S ribosome RNA gene [60]. The studies conducted in citation [61] and [62] suggests that LAMP exhibits high values of sensitivity and specificity for other species of Malaria. Being a more reliable and cost-effective method can easily replace PCR for screening in Malaria affecting regions. The only disadvantage is that the chemicals require cold storage facilities and further clinical trials are required [49].

1.3.4.3.3. A Flow Cytometry Method Assay

A Flow Cytometer is a device based on Coulter principle is a laser or impedance based cell counting and sorting machine where the cells are suspended on a solution stream. The use of such cytometer to detect haemozoin formed as a consequence of Malaria parasites digesting RBC was reported in [63]. When Haemozoin flows through cytometer depolarization of laser is detected. The performance analysis for the use of cytometry yields sensitivity value range of 49-98% and the range for specificity value of 82-97%, in case of Malaria detection [64]. Such methods are susceptible to error prone reporting of cases when there are different types of infections. Moreover, the systems are very costly and are unaffordable in under-developed countries.

1.3.4.3.4. Automated blood Cell Counters (ACC)

The Cell-Dyn is a popular cell counter that uses laser light multiple-angle polarized scatter for WBC separation and analysis. They have been found to detect haemozoin containing monocytes and granulocytes. In research citations [65], Cell-Dyn 3500 was used to detect Malaria infection. High sensitivity and specificity of 81.3 and 80.1% for Malaria infection [66]. The performance of Beckman Coulter in detecting haemozoin achieved a sensitivity value of 98% and specificity value of 94% [67]. The research citation [68] revealed that Malaria infection exhibited unusual distributions plots in the white cell channels.

1.3.4.3.5. DNA Micro Arrays

A DNA microarray is also known as biochip. A particular genomic expression or DNA is embedded on a region containing picomoles as probes or short fragments of DNA. The probe and target hybridization reaction can be detected as fluorophore. These microarrays are future diagnostic mechanisms [69]. Multiple probes can be embedded within a chip for detection of multiple conditions or targets. Such miniaturized chips can be automated for the development of microarray and falciparum Malaria has been detected in clinical specimens [70]. However, such diagnostic aid requires further development.

1.3.4.3.6. Mass Spectrophotometry

Mass Spectrometry (MS) is a technique that ionizes chemical and sorts these ions and a plot is obtained depending on their mass which is proportional to the charge. Such ion charges are obtained when any kind of matter is bombarded with electrons which gets distributed in an electrical/magnetic field based on the mass-to-charge. The mass of atoms and molecules are constant and known and are used to label by comparing the mass pattern. This method can be employed for detection of Malaria parasite in-vitro even at Parasitaemia levels of 10 parasite/µl. A direct ultraviolet Laser Desorption Mass Spectrometry (LDMS) is performed and for identifying Malaria parasite a specific biomarker Haemozoin is used. This method is also very fast with high output and can be automated for mass screening. Compared to conventional microscopy that requires 30-60 min, each sample takes less than a minute [71]. The only disadvantage is that it is difficult to implement in rural remote areas where there is absence of electricity.

1.3.4.4. Virtual Microscopy

Conventional microscopy utilizes normal light microscopy. The accuracy of detection is heavily dependent on the heuristic knowledge of the pathologists. Evaluation of slides by pathologists requires time and often prone to error of judgement. Other techniques like the use of RDTs though very popular and provide a rapid diagnosis, but the sensitivity is lower than microscopic evaluation. Moreover, they are specific for a particular species. Other techniques described require costly setup and equipment. Maintenance of such elaborate systems in remote locations is not possible. For efficient Malaria detection using the 'gold standard' required a technological upgradation. Advancement in microscopy, coupling them with a digital camera and connecting it with the computer system using firmware has ushered a digital era in the microscopic evaluation. Contemporary microscope does not use reflected light rather they have built in Light Emitting diode (LED) lighting systems with an array of light filters and fluorescent microscopy. Most of the problems associated with illumination in conventional microscopy are done away with the use of inbuilt illumination system. Modern digital camera with very high resolution can capture minute objects. Modern microscope systems also contain motorized stage for fully automated image capturing. Moreover in addition to this there are slide scanners or WSI that can capture the entire slide surface in a large image file. They mostly operate in batch processing mode which makes it suitable for mass analysis and hence lucrative to pathological industry. High resolution images generated by these scanners allow the pathologist to have the entire view of the slide on their computers instantly. With the development of microscope technology, computer software and algorithms are required for efficient identification of parasites. The development of CAD software for Malaria detection and diagnosis is one of such many applications of digital microscopy.

1.4. Research Objectives

The gross number of Malaria cases reported is increasing worldwide. There are several limitations with the traditional methods for Malaria diagnosis. Modern approach for disease diagnosis are accurate, however, implementation of such systems are difficult in poor countries. Moreover, remote areas are devoid of proper infrastructure which makes implementation of such systems impossible. The use of PBS examination of thick and thin film is widely accepted for detecting Malaria, species identification and for parasite count (Parasitaemia). The advent of virtual microscopy has provided the oportunity for further advancement of this technique. Development of a CAD system for Malaria disease detection and classification have been proposed using digitized microscopic/WSI images, detection. The objective of this work is to establish efficient, cost-effective and reliable algorithms to detect (as well as fast screening) the presence of two species of Plasmodium parasite, *P. vivax* and *P. falciparum* along with stage classification.

1.5. Contribution to the Knowledge Domain

The objective of the research work is to develop a completely automated Computer aided diagnostic system for Malaria. In this research work, the methods provide complete information to the pathologist for proper diagnosis. Information particularly related to the species of infection is vital from the clinical perspective. The knowledge of species and extent of infection (Parasitaemia) are vital for disease prognosis and management.

To develop such a system, we propose a hybrid (i.e. rule-based and machine learning based) algorithm. Such a hybrid system is most suitable to detect and analyse Malaria parasite. Simple rule-based methods have been used to detect suspect region and for feature extraction process. The suspect regions are further explored and features are extracted.

The feature list used for the purpose of classification are categorized into two groups, namely, Texture based statistical features and morphological features. The texture features were selected based on literature review. Apart from the morphological features pertaining to shape, size etc., we have introduced certain features in line with the pathologist's perspective. These features are particularly novel and unique for Malaria parasite identification. The concerned novel features are directly based on the characteristics of parasite morphology, cell deformity pertaining to the two species (i.e. *Falciparum* & *Vivax*) under consideration. These parasite morphology and modified cell morphology features are unique for this domain and are in coherence with pathologists' perspective in disease identification.

To classify the parasite, in terms of both, specie and stage of infection, widely used classifiers have been used. Ensemble classifiers and boosting algorithms are used to get the best classification results. The research work performed utilized several classification strategies and is documented in this thesis. A detailed study of classifiers was undertaken to select the best possible strategy. Simple supervised learners like k-NN, SVM and Naïve Bayes were selected. Similarly, it was observed that these classifiers when combined in an Ensemble provided better results in classification over individual use of these weak learners. A weight distribution was considered based on the individual performance of these weak learners and voting was performed to ascertain the classification label for a test datapoint. Adaboost was also considered for classification. This classifier is also based on the Ensemble principle.

Pathologists were consulted at every step to develop a system that abides to the standards documented by World Health Organization. The proposed research work is different from others in this respect.

The work contributes to the domain of research by proposing several new features that are specific to the disease. Similarly, new algorithms are proposed for fast screening of low resolution images. The use high resolution images for development new algorithm for pre-processing, segmentation, feature selection and classification to detect the presence of Malaria, stage and specie classification. Hybrid Method was used for parasite detection, parasite morphology and cell deformity identification for proper classification.

1.6. Organization of the Thesis

Chapter 1, provides an overview of the research topic at hand. This chapter provides a detailed insight on the Malaria disease, the parasite causing the disease and different diagnosis methods currently utilized by pathological laboratories. This chapter also defines the research problem at hand and also documents the key achievements of the research work conducted. Finally, the last section of the chapter defines the overall structure of the thesis.

Chapter 2, documents the literature review. Contribution to the digital pathology domain are limited as this knowledge domain is relatively a new concept. However, quite a number of authors have contributed to this area of research. A detailed review is performed on existing literature that will provide an insight towards the current scenario of research in this knowledge domain. The final section of this chapter identifies the loop holes or research gap. The objective of this research work is to overcome these research gap and to develop a robust system that will assist the pathologist.

Chapter 3, documents the overview of the proposed datamodel. It describes the building blocks of the proposed algorithms.

Chapter 4, describes the data that has been used for the research purpose. The data acquisition and other parameters of the data is described in this chapter. Information pertaining to all the three dataset are also documented.

INTRODUCTION

Chapter 5, details out the pre-processing performed on the images so that automated algorithms can use these images effectively. Pre-processing of images like illumination correction, background separation, artefact removal, noise removal, declumping of clumped cells, suspect region identification and Parasitaemia calculation is described in this chapter. The chapter documents both the initial screening, detection and classification phases.

Chapter 6, describes the different features that are extracted from the images in both the initial screening algorithm as well as the detection and classification algorithm. Feature selection and feature reduction is also discussed in this chapter.

Chapter 7, documents the classification methods implemented by both the screening and the main detection and identification of stage and specie module.

Chapter 8, documents the experimental results obtained for both the sreening and detection/classification phases. The chapter extinsively provides the detailed results for each step of algorithm execution. A thorough discussion and comparison with relevant work in this domain is also documented in this chapter.

Chapter 9, discusses the limitations faced by the proposed algorithms and tries to identify ways to overcome such limitations.

Chapter 10, compares the existing laboratory method with the proposed system in terms of execution time and cost effectiveness. This chapter was compiled with extensive consultation with the pathologist.

Chapter 11, summarises the achievements of this research work and also documents the future scope of this work in the digital pathology domain.

CHAPTER 2
LITERATURE REVIEW

2.1. Overview

The intervention of technology to assist pathologists and medical practitioners is vital for the fight against Malaria and its early diagnosis to prevent mortality. The biologists and the chemists were busy discovering means to control the disease. Advancement in microscopy and computer technology has bolstered this effort. Several works can be found in the research domain that have significant contribution. Several authors have tried to compare different methodology both in terms of detection technology and computational methods to establish better ways to identify and diagnose Malaria.

Research work related to the medical/clinical methods for Malaria parasite detection are dated to pre-historic times. However, research work on Malaria in context to medical image analysis is a relatively new concept. As already discussed with the advent of digital imaging the focus has shifted to image analysis as a method for detection of parasite within the thin/thick blood smear images. Quite a number of research work was focused on identifying Malaria parasites as well as the type of specie and lifecycle stages from image/s. Most of the research work conducted during the early years of digital pathology was to identify simple graphical software tools for better understanding of the images. Subsequently, the use of image processing algorithms for preprocessing and segmentation, as a part of image analysis, was also utilized for parasite identification. Most of these systems were based on rule-based image processing techniques. Image segmentation techniques like image binarization, edge detection and color-thresholding was widely used by authors for parasite segmentation. Morphological operators and Morphometry has also been effectively implemented for segmentation and parasite detection.

However, recent works have shown a shift towards machine learning approach with supervised and un-supervised learning approach in both image segmentation and parasite identification. Authors have derived host of features including color channels of different color model, textural and morphological features from the images. Using these features they have implemented supervised/ unsupervised machine learning algorithms to segment/classify parasite and cellular components. Some of the research works are focused on identifying the infection, stage and/or species directly from the slide images without segregating the infected region as region of interest. For development of an effective decision support system for the pathologists, the CAD system should be able to screen for infection, extract region of interest, segregate from normal cells, extract the parasite region, differentiate between species and stage, quantify the infection regions and calculate Parasitaemia.

A thorough review work has been undertaken prior to the actual research work. The objective of the review work was to identify the different research works carried out by authors in the particular domain. From the review work the different aspects related to the work to be conducted was identified. The need for a new system that envisages the complete spectrum while keeping it useful and manageable was documented. The review work was the basis for setting the research goals that the research work will try to accomplish. Contributions in this domain of research is discussed in this chapter of thesis.

2.2. Review Materials and Methods

At the initial stage of research work, a thorough review of existing works was performed. Locating the requisite articles and downloading documents was the first step in this direction. Documentation pertaining to Malaria research was available in the print media since the discovery of the disease by Laveran in 1880. However, present day most of the articles are available through e-versions of journals (both in medical and computer science domain). Some of the older literature/books/articles are also available in electronic media format and hence was easily accessible.

2.2.1. Article Crawling

The most convenient way to search and retrieve articles was to write proper search lines and using GoogleTM Search Engine. The other ways of obtaining relevant articles was to search medical indexing sites like PubMed Central or PMC, BioMed Central or BMC, Mendeley, Google Scholar, Thompson Reuters and many more. Search was conducted for materials with English text and the abstract was viewed before the article was downloaded. Review literature was critically analysed and references present in those articles were further investigated for relevant articles.

2.2.2. Selection of Articles

Articles collected for literature review comprised of journal research articles, literature review and survey, conference proceedings, books, book chapters, web-articles, webpages and reports. Research articles between years 2000 to 2018 were considered with an exception of one important article of 1966. All the materials acquired were categorized into different folders. A total of 291 documents were considered for the detailed study. These documents were required for general understanding of Malaria disease and contributions in terms of computer algorithms and methods. Out of the total 291 documents 237 documents (including research articles, conference proceedings and book chapters) pertaining to computer algorithms for Malaria detection and analysis. The 237 documents were further categorized based on the segment of image processing and/or document type (review/thesis/original research articles). Figure 4 depicts the different categories of documents acquired. Similarly, Figure 5 categorizes the different types of documents pertaining to computer algorithms on Malaria image analysis. The last figure (Figure 6) shows the frequency of papers on Malaria image analysis and CAD. However, not all of the research citations was considered for review work. A subset of these articles were selected based on their contribution to the knowledge domain.

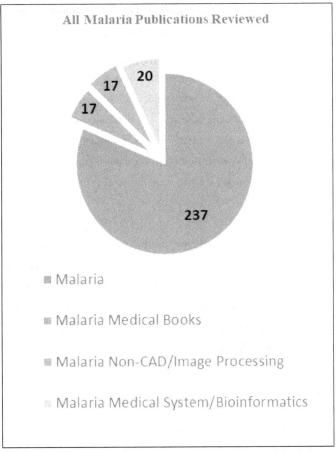

Figure 4. Total of 291 research articles on Malaria and related medical texts was investigated

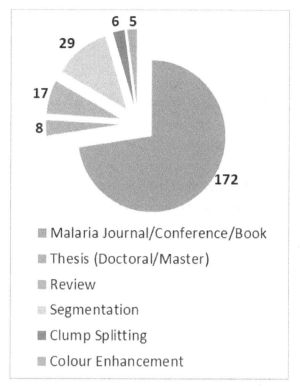

Figure 5. Distribution of the 237 research articles under consideration for the review work

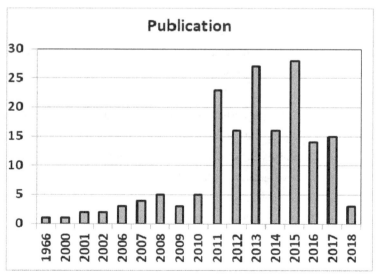

Figure 6. Distribution of Total 172 Papers including Journal, Conference Proceedings and Book Chapterswith reference to the publication year

2.3. Review Methods

A number of review articles can be found in the research domain that have tried to compare different methodology both in terms of detection technology and computational methods to establish better ways to identify and diagnose Malaria. The authors Tangpukdee et al. [72] studied different laboratory diagnostic techniques for Malaria detection including PBS. The authors reported the use of conventional microscopy for determination of Malaria by an expert based on visual examination of peripheral smears, as the 'gold standard for Malaria'. The study fundamentally compares different possibilities for the diagnosis of Malaria on a medical technology perspective rather than on a CAD based standpoint. Author Hanscheid T. [73], reviewed different alternative means to Malaria parasite detection other than thin and thick PBS method. He compared several modern molecular methods for determination of parasites from human blood sample. Similarly, authors Wongsrichanalai et al. [74], compared microscopy technique with RDT. The author performed a comprehensive survey of RDT kits available in the market.

The authors Tek et al. [75], performed a critical review of different algorithms of computer vision, image analysis and pattern recognition for automated Malaria screening using thin blood smears digitized images. According to the authors, most of the research work surveyed by them are a partial solution and cannot be used as a diagnostic aid. The main objective of the review work undertaken by the authors was to identify the advance methods employed, to provide an overall structure for further research and to determine different aspects of the problem for effective solution. The review article discussed different image processing methods that can be applied for parasite detection. The digital image acquisition methods, the variation in colour image arising due to different camera parameters and light source used, illumination variation, different thresholding problems associated with the image and solutions thereof. The authors also surveyed methods that performed detection process based on colour, morphological features, scale and Granulometry, cell size estimation, segmentation, and different classification methods including cross validation performed to remove training bias. Author Frean J. [76] have emphasized on Malaria load determination or Parasitaemia calculation based on medical methods and computerised algorithms. The authors established image processing approach as an alternative for parasite quantitation.

The authors Mohammed et al in the review citation [77] performed a detailed survey of classifiers in the machine learning domain that was utilized by different authors for parasite detection and classification. The authors discussed methods used under supervised and unsupervised machine learning techniques. The feature extraction methods like the use of PCA and the use of supervised and unsupervised ANN, SVM, k-Means clustering, Adaboost and Multiple Classifier Systems or MCS used at different levels of pattern classification, segmentation and pre-processing steps. The authors Chandra et al. in the review citation [78], surveyed several types of research works related to CAD based systems for Malaria parasite detection and also discussed about different diagnostic processes. In the review article by Jan et al. [79], the authors studied the methods proposed by other authors for Malaria detection from PBS slide images. The authors documented different image processing algorithms that have been commonly used by most authors. The authors compared the performances reported by research articles they have surveyed. Authors Rosado et al. [80] surveyed research articles on image analysis of PBS images. The authors differentiated the survey into two categories, namely, thick and thin smear images. The authors categorized the work into groups like segmentation, feature extraction, feature selection and classifications. The authors documented the variety of research work pertaining to these groups and also provided a critical discussion on the different research work reported by them. Mathematical Morphology and related features was surveyed by Loddo et al. [81]. The authors considered only those literature that utilized these morphological features for Malaria parasite detection the authors segmented the papers based on the pre-processing techniques utilized, segmentation of RBC and infection and feature extraction.

Authors Poostchi et al. [82], performed an extensive review of research work pertaining to the Malaria PBS image analysis with Machine Learning approach. The authors compared reported research citations and categorized the literature surveyed by them into image acquisition, pre-processing, red blood cell segmentation, feature extraction and selection, and parasite identification and labelling. The authors also surveyed the use of Mobile based CAD systems and use of Deep Learning in the Malaria image analysis domain. The review work presented by Devi et al. [83] has comprehensively discussed and documented different features used by authors for Malaria parasite detection from thin PBS images. The authors Das et al. [84] performed a systematic review of literature based on the different aspects of image processing. The authors categorized the survey of papers into groups namely, image pre-processing, illumination correction, noise removal, segmentation of RBC and chromatin region, microscopic feature extraction (morphological and textural features), feature selection, classification and CAD systems on Malaria detection. The authors Sadiq et al. in the review articles [85], [86] surveyed Machine Learning (ML) approach towards parasite detection. Similarly authors Jagtap et al [87] performed a general review by selecting a set of research citations. Authors Saputra et al. [88] considered *P. falciparum* in thin PBS images and reported works pertaining to this specification.

2.4. Digital Microscopy Based CAD Methods

The use of digital microscopy/ WSI scanners to obtain image data and further analysis by general purpose image processing software was the initial step towards automation. Subsequently, several scientists across the globe contributed towards the development of an automated disease diagnosis system that will detect the presence of parasites from the images obtained from digital slides. The search for the best possible method is still ongoing and remains a challenge in medical/computer science interdisciplinary research domain. This review work intends to investigate the efficacies and drawbacks of contemporary work done in this field. The authors have broadly categorized the research work cited here into two categories. The 'rule based system' category describes research work based on image processing and definite rules set by the respective authors to achieve the objective. In contrast, the second category, 'machine learning based hybrid systems' describes research work that utilizes image processing to obtain features to be applied to intelligent algorithms to train the system so that the system will derive its own rules to classify the image accordingly. Figure 7 shows the generalized scheme of Malaria parasite detection used by both category of authors.

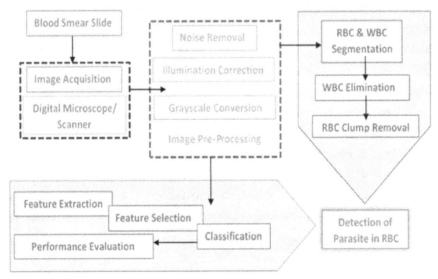

Figure 7. Generalized Scheme for detection of Malaria Parasite

2.4.1. Pre-processing

Pre-processing of digital images are important feature of any image processing algorithm. Images obtained from digital capturing equipments often suffer from illumination and noise related issues. Before apploication of any automated algorithm it is pertinent to make adjustment to the images in terms of illumination correction and noise removal. The illumination of images and colour density may vary intra/inter dataset. This arises due to different staining methods, stains and drying time. Moreover, the image capturing medium also utilises different lighting condition giving rise to illumination differences among images. Moreover color images vary based on staining methods used. For execution of an algorithm, it often requires some correction to maintain parity of conditions in the images. Tek et al [89] and [90], used image subtraction from a known image and low pass filtering whereas in the latter work they assumed the gray world normalization where they used a reference image. The images will have a constant gray value corresponding to a particular known image as the reference. Das et al. [91] adapted the 'Gray World assumption' as proposed by Lam et al. [92] for illumination correction. According to this approach the average of each channel is calculated. While keeping the green channel constant the gain in other two channels are computed.

To achieve noise elimination, Median filtering has been adapted by most authors like, Ruberto et al. [93], Ross et al. [94], Anggraini et al [95], Das et al. [91], Rosado et al. [96], Predanan et al [97], Bahendwar et al. [98], Makkapati et al. [99], Gitonga et al. [100] and Nugroho et al. [101]. Authors Dave et al. [102] and Savkare et al. [103] have used a combination of Median filtering with Laplacian filter for noise removal along with enhancement of the edge region. Adaptive and local histogram equalisation method is used for illumination correction by Sio et al [104], Purwar et al. [105] and Somasekar et al. [106]. Gaussian filter is employed by Arco et al. [107], Somasekar et al. [108] and SUSAN filter by Ahirwar et al. [109], Khan et al. [110] for noise reduction. The authors Reni et al. [111] performed contrast enhancement on grayscale image by finding optimum weights for R, G and B channels. Author Diaz et al. [112] performed low pass filtering for noise removal.

Morphological operations have also been used by authors for image pre-processing. The morphological operations of dialation and erosion helps in removal of unwanted artefacts and noise from the image. Such morphological operations are performed by several authors like Ross et al. [94], Tek et al. [89], Ruberto et al. [93] and [113], Khan et al. [114], Kareem et al. [115] and [116].

Automated cell clump removal is vital for accurate enumeration. The red blood cell de-clumping is performed by a rule based binary splitting algorithm as proposed by Kumar et al. [117] is also employed by Sio et al. [104] to de-clump red blood cells for accurate enumeration result. Diaz et al [112] used a template matching method by sliding a template over the clump region to separate the cells. Watershed transform method for clump splitting has been used by Preedanan et al. [97] and Bairagi et al. [118] (with Euclidian distance transform).

2.4.2. Rule Based Segmentation

Segmentation is the key for successful parasite detection. The process of segmentation is used to differentiate cellular components as as parasite and non-parasite regions. The authors Ruberto et al [93], Tek et al [90], Das et al. [91], Ahirwar et al. [109], Prasad et al [119], Dave et al. [102] and Savkare et al [103] have used Mathematical Morphology and/or Granulometry for determining the size of red blood cells/segmentation of foreground.

Image binarization using Otsu thresholding for image segmentation was performed by authors Das et al. [91], Ahirwar et al. [109], Anggraini et al. [95]. Mehrjou [120], Rosado et al. [96] and Savkare et al. [103]. Dave et al. [102] have used grayscale histogram with Kurtosis (to determine uni/bimodal histogram) and then performed Otsu thresholding. The authors Bairagi et al. [118] have use Otsu thresholding on RGB and HSV colour channels. Preedanan et al. [97] have adopted adaptive histogram thresholding for segmentation. The authors, Savkare et al [103], Mehrjou [120], in there research citations have implemented Watershed transform with distance transform for segmentation of red blood cells. An alternative method of Marker controlled Watershed transform is performed by author Das et al [91] and [121] and Khan et al. [114]. Authors Damahe et al. [93] used Zack thresholding method on the 'V' or 'value' component of HSV image for segmentation. Authors Purwar et al [105] have used Active Contour model for segmentation of red blood cells.

The citation by Sio et al. [104] proposed a software 'Malaria Count' for automated Parasitaemia count using light microscopy and using Giemsa stain thin blood smear slide images. The images were obtained using 100x oil immersion objective and digital camera. The image pre-processing included an Adaptive Histogram Equalization [122]. For proper counting of cells they were separated by implementing clump splitting method proposed in [117].

The citation by John A. Frean [123], studied the use of open-access software that will assist manual counting for diagnosis of Malaria. The primary objective of the author was to compensate the intra- and inter-observer variations through the use of computer automation. The author considered noise may be present due to presence of Howell-Jolly bodies, yeasts, artefacts and noise induced by low lighting conditions. The author adjusted the brightness and contrast of images for better thresholding results.

The authors Ghosh et al [124] implemented methods to enhance the image and filter out unnecessary regions so as to identify the presence of parasites. Laplacian Filter was used to to sharpen the edges. The image was converted to grayscale by the average of the components. A binary image was obtained by comparing empirically defined threshold value with the value obtained by subtracting the maximum grayscale value obtained from image histogram and inverted grayscale value. Morphological operation of closing was applied in which first dilation was done followed by erosion to remove small contour regions. A gradient operator was applied to identify rough regions depicted by presence of parasite over smooth regions.

The author Somasekar in his research citation [125] have proposed an optimization model based on mathematical linear programming model that defines an objective function using a set of decision parameters and a set of constraints imposed by the model. The proposed model reaches an optimized solution when a maximum or minimum value for the objective function was accomplished maintaining all the constraints of the model.

The authors Damahe et al. proposed a system [126] that identified parasites and counted the number of RBC in the image field. The proposed work focused their attention towards the identification of chromatin dots of infected RBC and differentiating them from Howell Jolly bodies that mimic infected RBC. In the first phase, the RGB image was converted to HSV color space. The saturation component was used to obtain the image histogram. This was used to determine the Zack threshold value [127] and an empirically obtained offset value was added to obtain the threshold value for image binarization. Sequential Edge Linking algorithm [128] was used to increase the accuracy of segmentation. Morphological operator was used for hole filling. Clustering was performed and contour boundary was obtained. Pseudo colouring was done on objects to count the number of RBC cells.

The authors Cecilia Di Ruberto et al [93] described a composite method to detect and classify Malaria parasites from Giemsa stained blood slides to obtain the Parasitaemia. Since the manual counting was tiresome, time consuming and often prone to error a computer based approach will hasten the process of Parasitaemia calculation as well as obtain the lifecycle stage of the parasite additionally. The authors identified three basic problems associated with the development of an automated system. The differentiation of Malaria infected RBC from WBC, cell segmentation and parasite stage determination were needed to be performed to solve the problem. For identification of parasite infected cells, the authors have used the methods proposed by them in citation [129] where they have used morphological operators using size and colour information obtained from the image. The authors performed cell segmentation by methods proposed by them in citation [113] where they used morphology and thresholding for segmentation. For classification of parasites, the authors have used two methods involving colour histogram similarity and morphological operators. Classification was done with the morphological thinning process to obtain 'skeletonizing' of the parasite region and they were differentiated based on the frequency of endpoints. Similarly, the colour similarity was done using HSV colour model where the colour histogram region of the parasite was compared with the histogram of the infected RBC region to classify parasite stage. The authors tested the proposed methods on images and compared the results with manual counting by experts providing good results.

The authors Kumar et al. in the research citation [130], Otsu threshold was calculated to obtain a binary image. After thresholding morphological hole filling was done using dilation operation. The contour image was given a pseudo-colour. The number of such coloured region was counted to obtain the number of RBC. The parasite infected cells were obtained in the same way by altering the thresholding value

Mehrjou et al. [120] proposed an automatic mechatronics system to detect Malaria while helping to overcome the difficulties of manual methods. Segmentation of RBC was done by adaptive thresholding, the value of which was calculated using the grayscale histogram. The dual peak of histogram representing the cellular and background regions was considered as a probability density function and was modelled with Gaussian function to obtain the threshold value for defining a binary image mask. The image was converted to HSV colour space and intensity parameter was extracted. The image was reconverted back to RGB and the green channel was extracted. Noise removal performed using a 3X3 median filter. Contrast enhancement performed by correcting histogram of tiles of fixed window size separately and then using bilinear interpolation to remove colour differences along tile boundaries. Cell segmentation was performed by removal of artefacts and RBC hole filling. By implementing flood fill algorithm to fill the background and engulfing any small areas of foreground the artefact was removed. Holes represent small background regions were engulfed as foreground regions. For identifying of cell clumps each foreground islands were cropped and the mask colour was swapped. A distance transform was applied on the image. The overlapping cells were identified based on the distance measure. For feature identification, different measures of RBC morphology of normal and infected were considered like relative size, the eccentricity of parasite infested RBC, smoothness of the cell margin, texture and relative colour of infected cells.

The authors Komagal et al in the research citation [131] proposed an automatic technique that consists of thresholding, grayscale conversion, image binarization, edge detection, object tracking and labelling. The authors used Sobel operator to detect the edges of the images. To detect parasites blob analysis was done on the binary image to detect the Malaria parasite.

The author P. T. Suradkar in her research citation [132] obtained binary image through thresholding and boundary of objects extracted after edge detection. Holes appearing within the RBC were filled using flood fill algorithm. A colour based thresholding for the colour channel was performed using static thresholding values obtained empirically by the author to differentiate and extract RBC cells. The RBC was counted and again a colour based thresholding was performed to identify infected RBC which by observation are pale in colour according to the author.

The authors Parkhi et al [133] proposed an automated parasite detection module based on linear programming. The objective was to identify, detect and classify parasites.

Ghate et al [134] proposed two different models for automatic Parasitaemia estimation algorithms According to the authors the results of most of the automated systems were poor due to lack of detection of cells along the border of images. The authors proposed a segmentation based algorithms. The images were converted to grayscale by averaging the colour components and the noise were removed by spatial averaging. Thresholding was performed for segmentation and Dilation was done to enlarge the cell boundaries and to fill the holes.

The research citation by Chakraborty et al [135] dealt with the problem of detecting Malaria parasite form thick smear images with an ill-defined circumference. The authors proposed two separate techniques for segmentation. The first technique involved the use of the morphological method. The slide image was pre-processed to grayscale and was followed by binarization using Zack thresholding. Before performing binarization Difference in Strength map or DIS map [136] for each pixel was obtained. Further, Morphological operation closing with initial dilation followed by erosion was performed. The relevant portion of the image was obtained based on the area of objects ranging between 0.0018% and 0.014% of total area. Segmentation was achieved using Euler number and the ratio of the black and white pixel for each object. The second technique used by the authors implemented on the image with HSV colour model where an empirically derived threshold segmented the background and the target region comprising Malaria infected cells. The Malaria affected RBC was segmented based on the porosity value given by Euler number. Finally, the authors have combined the two algorithms where the binarization and morphological operations were performed followed by segmentation using the HSV component thresholds derived experimentally and Euler number for final identification of Malaria infected cells.

The authors J.E. Arco et al in the research citation [107] proposed a new method based on image processing technique for enumeration of parasite infected RBC. The proposed method was divided into four stages. The pre-processing stage included Gaussian filtering for noise reduction followed by adaptive histogram equalization performed for contrast enhancement. This was followed by image binarization. This was followed by mathematical morphology operation of closing for hole removal and finally parasite enumeration. The adaptive histogram equalization was vital since the image foreground and background was often having image intensities very close to the neighborhood.

2.4.3. Rule Based Parasite Detection

Author Ruberto et al. [93] determined the presence of parasite by using two distinct methods. Firstly, morphological thinning process to obtain a skeleton of the ring parasite and next using colour histogram similarity measure to determine parasite region. Author Halim et al. [137], performed parasite detection using a Variance based approach and separately a Colour Based Co-occurrence Matrix based matching technique. Authors Tek et al [89] used RGB histogram and probability density function to determine parasite region. Toha & Ngah [138], calculated a threshold value to identify parasite region followed by calculating the Euclidean distance to differentiate between each parasite cluster. Makkapati et al [99], segmented chromatin regions by means of Otsu threshold method using HSV colour model and computed distance of red blood cell region and obtained chromatin regions to differentiate from nucleus of white blood cells. Damahe et al. [93] and Dave et al. [102], converted the image to HSV colour space, while Damahe et al. [93] performed thresholding on the 'S component histogram', Dave et al. [102] utilized the 'Hue channel' for parasite detection. The authors Ross et al. [94], computed two threshold values, for red blood cell and for parasite region from histogram of the green component of the RGB image. Ghosh et al [124], utilized gradient operator and thresholding of image with a computed value to locate parasite region. Fang et al [139], used Quaternion Fourier Transform (QFT) to obtain the amplitude and phase spectrum of image and the inverse Quaternion Fourier Transform to locate parasite region. Elter et al. [140], used green and blue channels for obtaining threshold value, followed by morphological Top-Hat to determine parasite region.

The authors in the research citation of Raviraja et al [141], have used dimension, shape and color for parasite detection. The authors further used size and shape characteristics of nuclei for parasite identification. The moments of a function can be applied to shape analysis is the basic principal of the research work. The region moments represent a normalized grey-level image function with a probability density of a 2-D random variable. This was described using statistical characteristic moments. The authors have assumed that any pixel value other than a zero value to represent as regions of a moment and every unique shape corresponds to a unique set of moments.

The authors Somasekar et al. in their research citation [108] implemented intensity based and use of Morphological operators to isolate the Malaria infected cells. The image was initially converted to a grayscale image and noise reduction was done using a 5X5 median filter. The authors justified the use of median filter over other filter for noise reduction as it can remove outliers while preserving the sharpness of the image. An intensity based extraction technique was employed by the authors to suppress the background and highlight the high intensity foreground region. Hole filling of cells were done using morphological operators and finally morphological erosion to isolate the infected cells.

2.4.4. Learning Based Segmentation and Parasite Detection

The authors Nugroho et al. [101] used K-NN classifier with 'S component' in HSV colour space for segmentation. Khan et al [142] performed clustering on the 'b' component of the image converted to Lab colour space for obtaining parasite region.

Tek et al. [89], used K-NN classifier with 20 classes to identify four species and four stage for each species and normal cases. Ross et al. [94], used geometric and texture features with Back Propagation Neural Network for classification of parasite infected red blood cell. Use of Artificial Neural Networks (ANN) for segmentation and parasite classification is used by Bahendwar et al. [32] (with RGB and HIS features) and Nugroho et al. [101] (with statistical and textural features). Authors Diaz et al. [112], used different histogram features with Support Vector Machine (SVM) and multilayer perceptron for classification. Khan et al. [114], used different textural features with Feed-forward Back Propagation Neural Network (FF-BPNN) for parasite identification. Anggraini et al [95], used Multilayer perceptron model for classification. The authors Das et al [91], Ghosh et al [124], used Bayesian and SVM classifier for detecting parasite region. Das et al [121], used texture based features with Multivariate Logistical Regression for identifying parasite in thin smear images. Authors Savkare et al [103] implemented SVM (with colour, texture and shape features) for classification of parasite and normal red blood. Similarly, Preedanan et al. [97] used Statistical features from grayscale image and SVM with RBF kernel for classification, while Bairagi et al. [118] employed Statistical and Texture based features with SVM classifier. Colour and Texture features with SVM classifier is used by Rosado et al. [96]. The authors Annaldas et al. [143] uses Energy feature using Grey Level Co-occurrence Matrices (GLCM), Statistical features and Phase of Image based feature with SVM and ANN classifier for parasite identification.

Diaz et al proposed a semi-automatic method for Parasitaemia quantification [112]. The authors proposed a method that not only counts the number of infected RBC but also determines the life stage of infection. The RGB colour space was transformed using YC_BC_R into chrominance and luminance channel. The mean values of regions were calculated and the obtained values were further smoothened using the moving window method. To segregate the erythrocytes, a labelling process was implemented using k-NN classifier and the normalized RGB colour information that labelled each pixel as erythrocyte or background. The binary image was converted into a hierarchy of foreground and background connected regions. The smaller foreground regions misrepresented as background were merged with the foreground to obtain the mask for the erythrocytes. The problems of overlapping erythrocytes were handled by a template matching algorithm. A normal erythrocyte shape was used to match segmented cell clumps. The search used a chain code representation of the clumped shape contour and Expectation-Maximization (EM) algorithm. For classification of erythrocytes the authors used machine learning technique. The feature was extracted by partitioning the image into a block of 64, 256 and 576 and generating histograms with r and g components. Saturation level histogram was generated with HSV colour space. Gray scale histogram and Tamura texture were generated. Three Tamura texture features with high correlation to human observation were considered and Sobel Histogram with 512 bins was obtained by the authors. But to reduce the number of features only mean, standard deviation, skewness, kurtosis and entropy values of five histogram were considered while reducing the number of features to 25 features. A two stage classification was implemented by the authors wherein the first classified infected or healthy erythrocyte. The infected ones were then fed across 3 learning models for each stage of disease and fourth learning model for the artefact. Any misclassification of the first stage was dealt by this fourth model while false negatives in the first stage were left for visual classification. The authors have used multilayer perceptron neural network and support vector machine non-linear classifier for evaluation. The SVM classifier was exhaustively trained using a radial basis function and polynomial function whereas Multi-Layered Perceptron (MLP) was tweaked by varying the number of hidden layers. The authors then performed a 10-fold cross validation by combining parameters to obtain the best set.

The authors Tek et al. [89] presented a new model for detection of Malaria parasite from thin smear blood slide images. The authors proposed a new binary parasite detection algorithm that modified K Nearest Neighbour (KNN) classifier and validated the performance using a

LITERATURE REVIEW

Bayesian method. The authors further performed three different classifications for detection of the parasite, stage of lifecycle and type of infection using a single multi-class classifier. The method proposed performs a five step image pre-processing for standardization of size, intensity, and colour of the cells and stained objects as proposed by the authors in the citation [90]. Illumination. colour correction and scale normalization were done. Segmentation was done using Rao's method [144] using the average cell area value and double thresholding to obtain a foreground mask. Colour normalization was performed by taking an average value of a known set of images and compared with the test image. The resulting image produced a white background. The model relies on the fact that RBC size is between 6 – 8 µm. The authors employed Area Granulometry based cell size estimation studies as performed in the citation. The area threshold value was calculated for the average cell area to obtain a pseudo-radius for the cell region. For feature extraction the proposed model extracted colour histogram quantized to 32 colours, local area Granulometry and six ratios as shape metrics were used. The system extracted 83 features as proposed by the authors in citation [90]. For a multiclass classification, the authors used K-NN having only the parameter K for adjustment, thus making the classification immune to repetitive tuning mistakes. The distance between the test points was based on a normalised version of city block distance metrics and a variant of the Canberra metric. Since the presence of parasite cell was fewer than normal cells hence an unbalanced approach of biased K-NN classifier was implemented. The authors performed a 20-class (detection, species and stage), 16-class (species and stage) and 4-class species and 4-class stage classification. The identification experiments were performed using hold-out and leave-one-out cross validation methods. For evaluation of proposed model Fisher Linear Discriminant (FLD) and the Back Propagation Neural Network (BPNN) classifiers were used with sufficient training and test data.

The research citation of Savkare et al. [145] calculated Parasitaemia of Malaria infected, Giemsa stained thin slide smear images that were captured using a digital camera. The authors have used median filtering for smoothening the colour image and Laplacian filter for edge detection. The image was converted to grayscale and global thresholding was done on the image. The image was binarized using Otsu thresholding [146], artefact removal and hole filling were performed using a median filter. The authors used watershed transform for overlapping cell separation. The authors further implemented selective elimination of cellular components based on the size of cells to isolate the RBC which was labelled by the system. The authors have selected features like colour, geometrical features like radius, perimeter, area, compactness metrics and statistical features like Skewness, Kurtosis, Energy and Standard

Deviation, that provided distinct variation between normal and infected cells to be used for training purpose. For classification purpose a linear SVM was proposed as a linear discriminant classifier. All the features were organized in decreasing order and only the most important features were selected for each pair of classes.

The authors Ahirwar et al in the research citation [109] proposed an automated Malaria detection and classification method from thin blood smear slide images using Artificial Neural Network (ANN). The morphological features of RBC have been used as features with Artificial Neural Network (ANN) for classification. The authors pre-processed the images using SUSAN approach (Smallest Univalve Segment assimilating Nucleus) [147]. This performed edge and corner detection with noise reduction yet preserving the structure of the objects. The authors claimed that this operator performance was better than the popular Canny operator in terms of quality of edges and speed of execution. Further, the authors employed Granulometry on the segmented image for determination of shape and size of the objects in the images. The RBC was differentiated using their size and eccentricity. The algorithm proposed was based on the selection of two thresholding values obtained from the bimodal distribution of the green component. The first threshold separated the RBC from the background while the second separated the infection with affected RBC. The mask image was dilated to accommodate the parasite infected RBC. The artefacts like the nucleus of WBC was removed by the size that was obtained empirically a priori. The infected cells were isolated by reconstruction of cells using the mask. After determining the presence of Malaria parasite, the feature set was generated for the classification system. Two feature set was generated; one based on the morphological characteristics, colour attributes and texture and the other was the a priori knowledge of different measures of the parasite. The presence of infected RBC was established using Back Propagation Feed Forward (BFF) neural network.. The colour and texture of the parasites were used as training features.

The authors Nasir et al in the research citation [148] proposed a method for detection of Malaria parasites from images of a thin blood smear. A contrast enhancement method called Partial Contrast Stretching technique and the images were transformed into HSI model. The parameters of HSI colour model were used as features for the unsupervised clustering using Moving K-Means (MKM) for segmentation of parasite infected RBC.

The citation of Abdul Nasir et al. [149] performed parasite detection using different colour models and unsupervised learning. The authors extracted colour channel information of three popular colour models namely, RGB, HIS and C-Y as features to be used with k-means clustering algorithm. Further unsupervised pixel segmentation was performed using k-means clustering for segmentation and to obtain only the infected cells. The authors have empirically determined that regions with area less than 5000 pixels can be suppressed as not-infected. Shows the output images obtained by the proposed method.

The authors Suryawanshi et al. [150] in the research citation compared the performance of two classifiers for effective identification of Malaria parasite from digitized images of thin blood smear stained with Giemsa dye. A Multiscale Laplacian of Gaussian filter (LoG) assigned a marker to extract each cell from a cluster of cells and performs edge detection. A binary object labelling technique was added to extract the cellular dimension feature from the seed markers. Another feature was obtained using Gabor filtering process. The authors trained two classifiers, Euclidean distance classifier and Support Vector Machine (SVM) classifier.

The authors Kurer et al in the citation [151] proposed a method that performed Canny Dilation of the image followed by training using Probabilistic Neural Network (PNN).

The authors Chayadevi et al. [152] utilized machine learning models using colour parameters and fractal features. The image was transformed to grayscale, HSV and LAB colour formats.

The author, Razzak in his research citation [153] proposed an ANN based classification model for Malaria segmentation and detection. To classify and segment the Malaria infected cells, the authors performed segmentation and BPNN for classification. The author also obtained GLCM based 28 texture features. To classify Malaria infected cell the authors used Back Propagation Artificial Neural Networks.

Bahendwar et al [98] proposed a segmentation algorithm for Malaria parasite isolation based on Artificial Neural Network (ANN). The authors observed that the green colour component of RGB image as the best feature for segmentation and RGB features excelled over the other set containing HSI features with RGB.

There are other notable research citations that was recently published, that of Rajaraman et al [154], Gopakumar et al [155], Rosado et al [156], Bibin et al [157], Devi et al [158], Park et al [159] and Widodo et al [160] have made significant contribution to the domain.

2.5. Research Gap Analysis

The review work has been conducted prior to the actual research work undertaken. The objective of the review work was to identify the scope of further research in the domain.

The research work conducted in this domain can be categorized into two groups. One group of authors have used morphological operations and/or colour features for segmentation of cellular components after performing initial pre-processing to standardise the dataset. After segmentation the authors have proposed some rule-based systems to detect the presence of parasite within the RBC. Most authors have used some threshold-based method to establish the presence Malaria infection. The other group of authors have derived host of features including colour channels of different colour model, textural and morphological features from the images. Using these features they have used supervised/ unsupervised machine learning algorithms to segment/classify parasite and cellular components. Most of the research work focussed on identifying the infection, stage and/or species directly from the slide images without segregating the infected region as region of interest. Thus the image algorithms are implemented on the entire image leading to high execution time. Moreover, few methods attempts to identify parasite stage and species. For development of an effective decision support system for the pathologists, the CAD system should be able to screen for infection, extract region of interest, segregate from normal cells, extract the parasite region, differentiate between species and stage, quantify the infection regions and calculate Parasitaemia. Such a system is lacking in the digital pathology domain for Peripheral Blood Smear (PBS) images.

The infected regions are extracted as objects of interest. Morphological and texture features are extracted from the objects for pattern matching. Supervised learning algorithms are used for classification.

In a nutshell, the proposed system suggests a methodology for morphological and texture based feature extraction and thereby parasite and stage, specie classification using a hybrid (rule and machine learning based) algorithm. The Figure 8 shows the Gap Analysis and the proposed methods to mend the loopholes present in the current system.

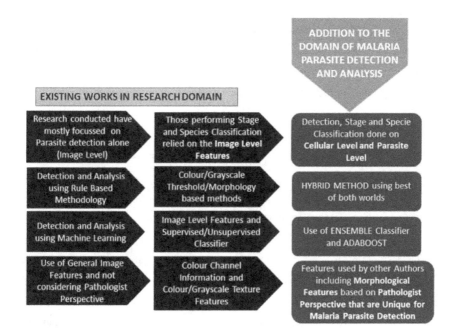

Figure 8. Gap Analysis as an outcome of thorough literature review. The poosible ways to mend the gaps present in the current system

2.6. Summary

Application of Digital Pathology for Malaria disease analysis is a relatively new concept. A few number of research works are present for identifying Malaria parasite as well as the type of specie and lifecycle stage from image/s. These research works used image-processing algorithms for pre-processing and segmentation, mostly based on rule-based image processing techniques. However, recent research works use machine learning approach based on supervised and un-supervised techniques. Features derived from the image with supervised/ unsupervised machine learning algorithms segment/classify parasite and cellular components. The review process identifies relevant literature present in the domain. The research documents were thoroughly studied and reviews already available were duly represented. A new perspective towards segregating the existing literature was adopted in this thesis. Existing works were differentiated based on their approach towards solving the research problem, namely, the rule-based and automated learning systems. A summary of relevant works pertaining to the two distinct approaches has been documented. The scope for further research was established after thorough study of the existing works.

LITERATURE REVIEW

Several research works were reviewed to highlight the different image processing techniques used by fellow scientists across the world. Modern machine learning models provide a better outcome and are being widely used. Methods applying such intelligent algorithms have also been discussed in this chapter. It may be concluded that application of image processing algorithms using computerized systems will enhance the field of diagnostic pathology. Computer as a tool for medical diagnosis will improve the quality of diagnosis by eliminating any human bias associated with decision making. Similarly, huge workload of pathologists can be greatly reduced by using computers with intelligent software tools. Advent of network centric models has decentralized the process of medical evaluation. The medical Practitioner may not be required to visit the remote regions affected by the disease. Images or samples taken by a technician can be remotely viewed by the concerned medical specialist. Digital pathology is a nascent and a promising area in the field of modern medicine. The use of digitized medical/diagnostic samples are easier to handle can be archived and analysed remotely by a pathologist. CAD tools can be applied to digitized images to assist the pathologist. The review chapter will provide adequate information for any research work in the field of CAD development for Malaria detection. Any future work in the area of Malaria detection can take into consideration the advantages/disadvantages of algorithms already present in this particular research domain. The use of hybrid systems, combining the best of image processing and machine intelligence, may lead to the development of future CAD systems that will have higher accuracy in Malaria detection.

CHAPTER 3
OVERVIEW OF DATA MODEL

In view of the identified loopholes, a model was developed towards Parasitaemia detection along with stage and specie classification from thin blood smear image. This chapter has two sub divisions, while one provides a block based representation of the proposed data model (i.e. block representation of the data model), the other section is aimed to provide an overview of each of the parts that contributed towards knitting together of the proposed model (i.e. a brief description of the proposed model). The work proposes a hybrid method for Malaria parasite identification and subsequent stage–specie classification. The model developed is a hybrid as it is a strategic amalgamation of rule based logic and machine learning algorithm.

3.1. Block Representation of the Data Model

The objective of the research work is to develop a completely automated computer aided diagnostic system for Malaria. In this research work, the methods provide complete information to the pathologist for proper diagnosis. Information particularly related to the specie of infection is vital from the clinical perspective. The knowledge of species and extent of infection (Parasitaemia) are vital for disease prognosis and management. To develop such a system, a hybrid (i.e. rule-based and machine learning based) algorithm has been proposed. Such a hybrid system is most suitable

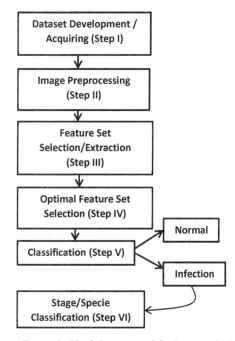

Figure 9. Block Diagram of the Proposed Model

to detect and analyse Malaria parasite. Simple rule-based methods have been used to detect suspect region and for feature extraction process. The suspect regions were further explored and features were thereby extracted.

The feature list used for the purpose of classification were categorized into two groups, namely, Texture based statistical features and morphological features. The texture features were selected based on literature review. Apart from the morphological features pertaining to shape, size etc., certain features in line with the pathologist's perspective were also introduced and thereby included in model development. These features are particularly novel and unique for Malaria parasite identification. The concerned novel features are directly based on the characteristics of parasite morphology, cell deformity pertaining to the two species (i.e. *Falciparum* & *Vivax*) under consideration. These parasite morphology and modified cell morphology features are unique for this domain and are in coherence with pathologists' perspective in disease identification.

To classify the parasite, in terms of both, specie and stage of infection, widely popular classifiers have been used. Ensemble classifiers and boosting algorithms are used to get the best classification results. The research work performed, utilized several classification strategies that are briefly documented in this manuscript. A detailed study of classifiers was undertaken to select the best possible strategy. Simple supervised learners like k-NN, SVM and Naïve Bayes were selected. Similarly, it was observed that these classifiers when combined in an Ensemble provided better results in classification over individual use of these weak learners. A weight distribution was considered based on the individual performance of these weak learners and voting was performed to ascertain the classification label for a test data point. Adaboost was also considered for classification. This classifier is also based on the Ensemble principle.

Figure 9 provides a block based representation of the hybrid methodology that has been proposed for Malaria parasite detection. Each block of the model has been singularly and explicitly dealt with in the subsequent chapters.

3.2. A Brief Outline of the Proposed Model

The first section of the proposed model deals with initial screening phase. At this phase, thin blood smear images at 40X were used for initial screening to determine whether the thin blood smear image is Malaria infected or not. Based on the findings of this segment a probability is assigned to each image based on its chance to contain infection. Given the variability that encompasses Malaria parasite detection and the dearth of information in 40X thin blood smear images, if any image is found to have even 1% of the pixels as infected then the image is marked to contain Malaria infection and the corresponding 100X image is further intricately scrutinized for Malaria infection detection. If an image is found to have no infected pixel as per the screening algorithm then the image is considered normal.

Figure 10 provides a flowchart representing the working of the system for Malaria parasite detection. So in particular the system has 2 broad phases, namely the Initial 40X based screening phase and the main 100X based stage and specie detection phase interlinked based on infection percentage. The order of the subsections for each of the two broad phases has been duly represented in Figure 9.

The main detailed phase identifies as also classifies Malaria infection in terms of stage and specie. Though the main phase can be independently used for parasite detection and subsequent stage/specie classification yet it is made dependent on the initial system to enhance system efficiency in terms of time complexity from the perspective of a computer scientist but such efficiently wasn't found to hold much significance statistically as also within the clinical setting. The broad phases can be considered as the outer shells that house the algorithms pertaining to the subsections highlighted in Figure 9. These algorithms form the actual crux of this research initiative.

To test the viability of the system as a whole, performance analysis in terms of accuracy, sensitivity & specificity was performed alongside cost viability analysis for Malaria parasite detection in real world slide image.

In coherence with the flowchart outlined in Figure 9, Chapter 4 describes the datasets that were used for development and testing of the algorithms developed that contribute to the Malaria parasite detection system as a whole.

OVERVIEW OF DATA MODEL

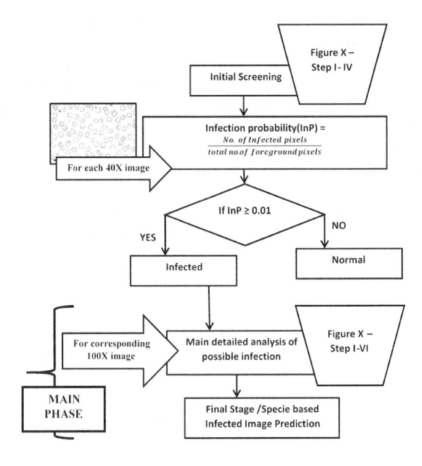

Figure 10. Flowchart of the Parasite Detection System

CHAPTER 4
DATA DESCRIPTOR

4.1. Overview

The datasets used for the study are three in number, namely, a publicly available dataset [161], dataset acquired from Government hospital in Kolkata with due ethical formalities and a mixed dataset obtained by combining the two. In terms of comparative analysis of the different datasets that have been used by other researchers working on Malaria parasite detection it might be integral to mention that most research studies have been conducted on single dataset of thin blood smear images (either images downloaded from the internet or custom dataset acquired from hospitals). In the research initiative under discussion, a total of 80 patient slides with 2730 images have been used. This dataset is more than the average number of dataset images (i.e. data points) that have been used by other authors/researchers in this domain. In coherence with statistical inference drawn for medical dataset analysis [162] [163] it can therefore be inferred that the number of images were sufficient to establish the findings of the study statistically thereby making the conclusions drawn on their basis acceptable within the scientific community.

4.2. Dataset Preparation & Sampling

Given that the Initial Screening phase was developed as a quick precursor to the main phase, the datasets used for the development and thereby testing of the Initial Screening phase and the main phase were particularly different. Owing to the fact, the the main phase was built on a larger dataset as opposed to the 250 image based MaMic database that was used for the initial screening phase, the algorithm developed for the main phase was more robust. Again, as abiding by WHO standard the main phase uses 100X magnification of images for Malaria parasite detection and subsequent staage/specie classification, this phase is considered reliable within clinical setting as opposed to the initial screening phase that uses 40X magnified images for Malaria parasite detection.

4.2.1. Initial Screening

For the Initial Screening phase thin blood smear images were acquired from the public MaMic database [161]. Images were acquired at 40X magnification and 25 watt illumination. Once acquired, image noise was corrected. The dataset used for developing and testing the algorithm at the initial stage consisted of 250 images. Of the 250 images acquired, 125 images consisted of blood infected with Malaria parasite, while the other images consisted of blood smears taken from individuals not infected with Malaria. Each image in the dataset was of size 1387 x 932 pixel^2. The size of the dataset was determined based on the Cochrane Q test. The images were selected based on Convenience Sampling done by the concerned Pathologist.

Once tested, the initial screening phase was integrated with the main method. It was used for all images(i.e. MaMic Database & Hospital Acquired images) in the main method at 40X magnification.

4.2.2. Detection and Classification Method

The glass slide of a thin smear of blood contains a spread of vascular tissue of an individual probably containing Malaria infection. For observation and analysing a digitised image requires the obtained image to be captured at 60X to 100X magnification. For the purpose of species and life-cycle stage classification 100X magnification is the standard. For each patient a single slide is prepared however as per WHO standards more than 25 non-overlapping images are required to be observed to report the absence of infection. Similarly, in presence of infection at least 300 RBC are needed to be observed to calculate and report Parasitaemia. Figure 11 shows sample images of Dataset #1 and Dataset #2. Table 4 shows the image specifications and dataset details.

4.2.2.1. Dataset #1

The database that was acquired from MaMic [161] (which is a publicly available database) pertains to snapshots taken from a whole thin blood smear slide scanned at 100X resolution of *P. falciparum* infection. Some of the infected slide scan areas were devoid of infection. This is due to low Parasitaemia for those slide scans. Some of these slides were taken into consideration to test the robustness of the proposed system.

4.2.2.2. Dataset #2

The dataset is acquired from the Pathology Department of a Government Hospital in Kolkata under the supervision of Dr. D. Ckakraborty. The slides were prepared by the Ronald Ross Malaria Centre within the hospital campus.

4.2.2.3. Dataset #3

A combined dataset of the images obtained from MaMic [161] and Government Hospital in Kolkata.

Specification	Value
Dataset #1	
Total Number of Scanned Slides	54 slides
Sample size based on Cochrane's sample size selection for small datasets	47 slides (23 Normal and 24 infected) 12 P.vivax/ 12 P.falciparum
Number of non-overlapping blocks from each slide	30 [Convenience based sampling][From a set of 2790 images] (Understanding Power and Rules of Thumb for Determining Sample Sizes)(each of size 5.08 x 3.39 cm^2 [600 x 400 pixel^2])
Image Resolution	300 dpi
Magnification used for each Digital Image	100X
Total Number of images used for the study	30 x 47 = 1410 images
Normal Images	743 (Some slide with infection have normal images due to low Parasitaemia)
Images with Infection	667
P.vivax/P. falciparum	330/337
Vivax Ring – Tropozoite - Schizont – Gametocyte	162 - 115 – 34 – 35
Falciparum Ring – Tropozoite - Schizont – Gametocyte	373 – 38 – 28 – 76
Dataset #2	
Total Number of Scanned Slides	33 slides (10 Normal and 23 Infected) 11 P.vivax/ 12 P.falciparum
Number of non-overlapping blocks from each slide	40 (each of size 5.5 x 2.96 cm^2 [650 x 350 pixel^2])
Image Resolution	300 dpi
Magnification used for each Digital Image	100X
Total Number of images used for the study	40 x 33 = 1320 images
Normal Images	400
Images with Infection	920

P.vivax/P. falciparum	440/480
Vivax Ring – Tropozoite - Schizont – Gametocyte	216 - 153 – 46 – 52
Falciparum Ring – Tropozoite - Schizont – Gametocyte	532 – 54 – 39 – 108
Dataset #3	
Total Number of Scanned Slides	80 slides (33-normal, 23-P.vivax, 24-P.falciparum)
Image Resolution	300 dpi
Magnification used for each Digital Image	100X
Total Number of images used for the study	1410 + 1320 = 2730 images
Normal Images	743 + 400 = 1143 images
Images with Infection	667 + 920 = 1587 images
P.vivax/P. falciparum	770/817
Vivax Ring – Tropozoite - Schizont – Gametocyte	378 – 268 – 80 – 87
Falciparum Ring – Tropozoite - Schizont – Gametocyte	905 – 82 – 67 – 184

Table 4. The parameters followed for dataset development with the description of different parasites as observed by experts (ground truth)

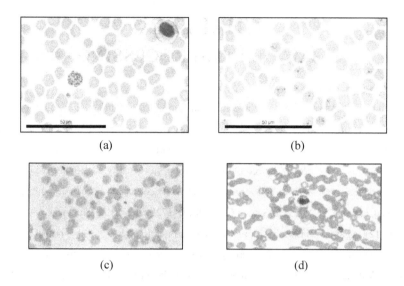

Figure 11. Sample Dataset at 1000X (a) MaMic image showing mature P. vivax Schizont, (b) MaMic image showing multiple infection of P. falciparum rings, (c) Hospital supplied slide image showing P. vivax Gametocyte and (d) Hospital supplied slide image showing P. falciparum Schizont

4.2.3. Dataset Annotation

Once acquired the dataset was duly annotated by Dr. Chakraborty and his team. The annotations provided by Dr. Chakraborty and his team was used at the ground truth in the development and thereby evaluation of the algorithm developed.

4.2.4. Dataset Standardization

All images considered were in bitmap format(.bmp) and were standardized to contain values in the range of 0 and 1 in order to remove overshadowing of smaller values by larger ones. Additionally, all images were read into Matlab 2017a in batch mode for development of the proposed hybrid algorithm.

To prevent clustering of similar images, image vectors (i.e. datapoints) were randomly shuffled for both the initial screening and main algorithm development phase during development of the data model.

4.3. Summary

Data set selection/development is crucial for algorithm development. Having described the dataset the next chapter deals with the process of separation of the foreground (i.e. the cellular materials) from the background.

CHAPTER 5
IMAGE PRE-PROCESSING

5.1. Overview

For both the sections and for both the concerned datasets, the staining colour for the images as also illumination for the images was found to vary in both inter and intra database images. Again intra cellular color variation for RBC cell in a particular image was also recorded. Owing to the oval, discoid shape of the red blood corpuscle, statistically significant colour variation was recorded for a particular cell represented in RGB colour space (Mann Whitney U-1336 (for Red Channel), U-305.5 (for Green Channel), U-1720.5 (for Blue Channel), p=0.000<0.01) (Figure 12). Again, variation in colour was detected within an image as a whole. Colour-based 3-means clustering was used to identify three colour clusters in the foreground pixels. A Kruskal Wallis test revealed that the median value of the three clusters is unequal (Red Channel $\chi^2(2)$=11927.094, p=0.000<0.01, Green Channel $\chi^2(2)$=12118.618, p=0.000<0.01, Blue Channel $\chi^2(2)$=11372.600, p=0.000<0.01). In addition to the statistically significant difference recorded in a particular image, significant difference in colour distribution was also recorded among two different slide images in the dataset developed (Figure 12) (Mann Whitney U-test p=0.000<0.01).

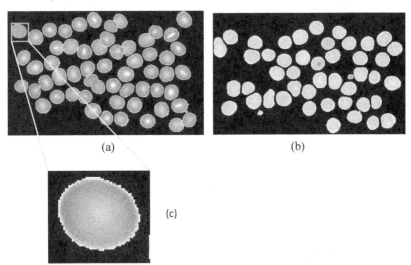

Figure 12. (a) and (b): Two sample images from the dataset (foreground extracted image) showing the inter image colour variations, each having a particular set of unique colours which are statistically significanlly different.(c) An enlarged view of a red blood corpuscle elucidating intra-cellular colour variation.Image showing intra and inter dataset variatios

To handle this discrepancy, illumination correction was performed as a precursor for either of the two processing blocks, namely, preprocessing pertaining to initial screening phase and subsequent main proposed system.

5.2. Illumination Correction

Illumination correction has been implemented by Diaz et al. [112], Tek et al.[89]. A Kruskal Wallis H test performed on a randomly selected set of three images revealed that there is significant difference in luminance distribution in the images, $\chi^2(2)=21.706$, p=0.000<0.01. Thereby, suggesting significant difference in luminance among images in the MaMic dataset. For the datasets under consideration, automated luminance correction of image is performed to extend the applicability of the algorithm across multiple datasets. Figure 13 depicts an image from the MaMic dataset whose luminance has been particularly marred to test the performance of the algorithm in question. A single or the same image was used as the reference luminance for all images in the dataset. In Figure 13, for each test image (in this case Figure 13a) the luminance matrix of image 13b was used as reference. The ratio of the difference in the standard deviation of the test (σ_{test}) and reference image ((σ_{ref}) against the standard deviation for the reference image was used to increase or decrease the luminosity of the RGB image under consideration.

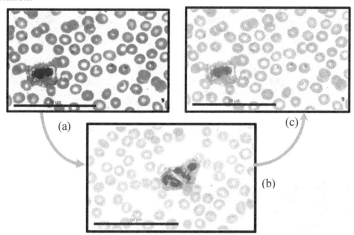

Figure 13. (a) An image whose luminance has been degraded on purpose to test whether the algorithm works on other image datasets with bad luminance.[Luminance distribution specification provided in Figure 14].(b) A reference image from the MaMic database to improve the illumination of image (c) Luminance improved image.

IMAGE PRE-PROCESSING

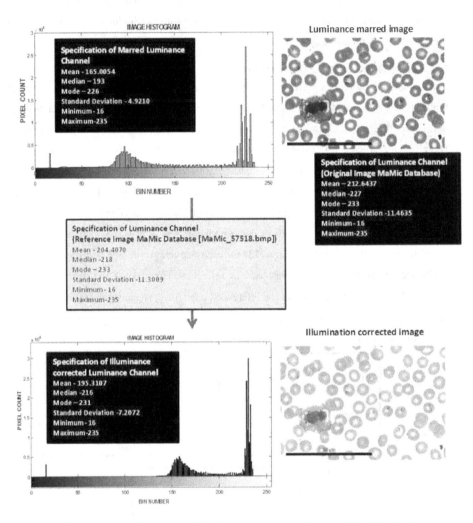

Figure 14. Represents the histogram of the Luminance marred image of the MaMic Database (original Image MaMic_57503.bmp) along with the original specifications for the same. Additionally, the specification for the luminance channel of the reference image and the final illuminated image along with the luminance channel distribution specification have also been documented in the image.

It must be noted that, even such dynamic illumination correction of image based on a reference image does not particularly work in case of a differently stained digitized thin smear image. So for each dataset differing from the other in terms of colour composition, the illumination correction phase had to be customized using a reference image from the concerned database.

Thus the illumination correction phase was particularly bypassed to develop an algorithm encompassing differently coloured/stained digitized thin blood smear images.

5.3. For Initial Screening

To correct salt pepper noise 2D Median filtering with 3 by 3 window was performed. Once noise corrected, the RGB images were converted to Lab Colour Space image. Based on the a and b components, unsupervised K-means clustering was performed to segment out Red Blood Cells from the Geimsa stained thin blood smear images. The methodology has been duly represented in Figure 15.

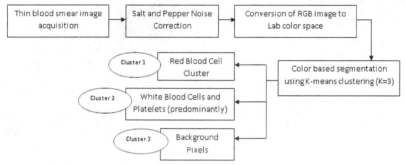

Figure 15. Block diagram depicting the process of segemntation of the image to obtain the foreground region.

Based on the a and b components, unsupervised K-means clustering was performed to segment out Red Blood Cells from the Geimsa stained thin blood smear images. Figure 16 illustrates the cluster structures formed based on the pixel values of the image represented in Lab colour space.

IMAGE PRE-PROCESSING

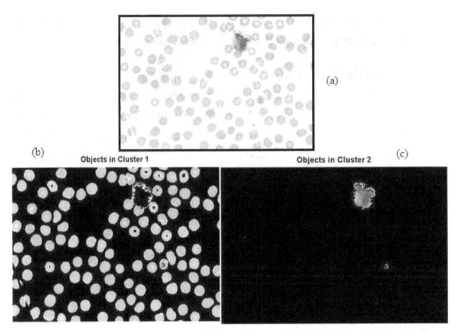

Figure 16. (a) Thin Blood Smear Image, (b) Cluster 1 consisting of Red Blood Cells, (c) Cluster 2 consisting of White Blood Cells, Platelets and Malarial Parasite.

Based on empirical work as also experience, K-means clustering algorithm was designed to produce 3 clusters (i.e., K=3). While cluster 3 consisted only of the background pixels, cluster 1 consisted of red blood cells and cluster 2 consisted of white blood cells and platelets. Though clusters are particularly for naming purposes named as cluster 1, 2, and 3. Based on the K-means clustering algorithms, often the components of the named Cluster 2 were represented as Cluster 1 or 3 and likewise. Cluster 2 and Cluster 1 as named were identified automatically by the proposed system based on the number of connected components and the average area of the connected component as represented in the clusters.

The red blood cell cluster(i.e. Cluster 1) was binarized. The Sobel operator was used to detect the edges of the red blood cells. The image was eroded emulating a disc structure of radius 3 and neighbourhood size 4. The number of connected components was enumerated. Disc based erosion was performed as red blood cells have a disc like structure (refer Figure 17). Even for thin blood smears, clumping of red blood cells is a common occurrence that makes enumeration of red blood cells complicated. In coherence with the work of Tek et al [89], the red blood cells were de-clumped and counted, to obtain a value of the total number of red blood cells.

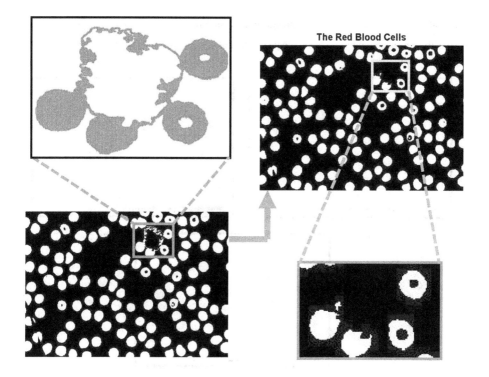

Figure 17. Disc Shaped Erosion was performed to remove trace of WBC outline within the RBC cluster

The WBC outline within the RBC cluster was thereby removed (Figure. 17) using Disc Shaped Erosion

5.4. Main Phase

In the Main Phase of the Proposed System, the foreground connected components were segmented out from the background pixels. The foreground components were thereby represented in YC_bC_r colour space to reduce the number of distinct colours and thereby recoloured to segment out WBC nucleus material, possible infection suspect regions for further investigation.

5.4.1. Background Separation

This research work defines background as any pixel that is not part of the cellular components of the thin blood smear image. Background separation is integral towards labelling of the cellular components while negating out the spurious pixels in the blood serum along with the unwanted image artefacts, namely, the reference scale. Figure 18 represents the non-uniform distribution of colored background for an image in the database along with the dimension

IMAGE PRE-PROCESSING

reference artefacts. The method of background separation can particularly be divided into 2 basic sub-sections, namely, reference scale artefact removal and non-uniform background pixel removal.

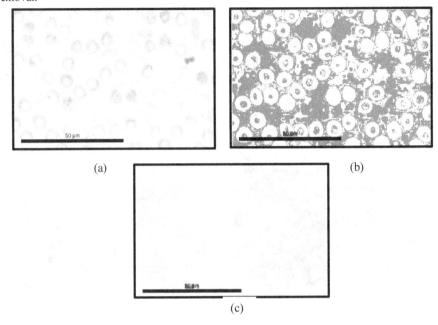

Figure 18. Sample image from MaMic database, (a) Input image (RGB) with background, (b) the image in RGB colour space. (c) The foreground identification that was used as reference point (ground truth) for evaluation of the performance of extraction algorithms

5.4.2. Reference Scale Artefact Removal

Given that the scale has uniform black colouration, rule based colour pixel recolouring was performed to remove the scale label markings having multi-colouration and being located at the same position for each of the images in the MaMic image database, rule based positional recolouring of pixels was performed. The scanned hospital image database lacked the scale artefact.

5.4.3. Non-uniform Background Pixel Removal

Two different approach based methods were compared to separate out the background from the cellular components in the thin blood smear image, namely, colour based k-means clustering and image histogram based modified Zack's thresholding [127] (Figure 19b) was used for removing the background of the image.

Three cluster based k-means (k=3) clustering was performed on the image represented in the L*a*b colour space without applying any image correction techniques. For the particular image clustering was performed using squared Euclidean distance among the pixel values represented in L*a*b colour space (Fig. 19a).

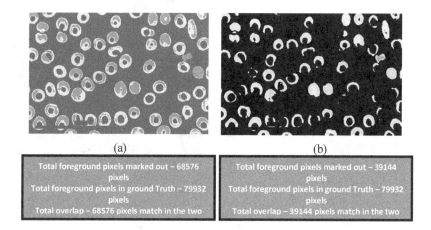

(a)   (b)

Total foreground pixels marked out – 68576 pixels Total foreground pixels in ground Truth – 79932 pixels Total overlap – 68576 pixels match in the two	Total foreground pixels marked out – 39144 pixels Total foreground pixels in ground Truth – 79932 pixels Total overlap – 39144 pixels match in the two

Figure 19. Output image for automatic foreground extraction (a) using 3-means clustering (b) Output obtained for 3 threshold based Zack's clustering algorithm based on the 3 most prominent peaks in the indexed image histogram.

With reference to the sample image represented in Figure 19, 3-Means algorithm was used as the basis for the subsequent steps in the methodology section.

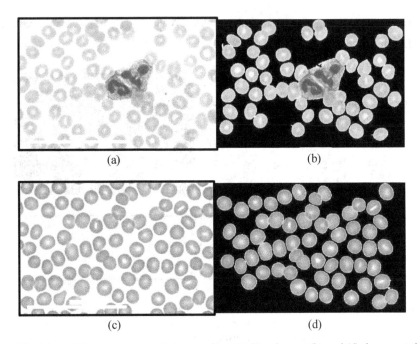

Figure 20. (a) and (c) represent sample images from MaMic dataset (b) and (d) document the extracted foreground cellular particles after suppression of non-uniform background. As per WHO guidelines for Malaria Microscopy, the cells in the border region were removed.

As demonstrated in Figure 20(b) and 20(d), the Red Blood Corpuscles to the periphery of the image for either of the images have been eliminated as per the guidelines set out by WHO for Parasitaemia estimation.

5.4.4. Image Recoloring

To enable broad categorization of the different RGB based colour values, the RGB image was translated into the YC_bC_r colour space. This helps to down-sample the coloured foreground image. Based on the YC_bC_r values the image is automatically recoloured based on the largest of the 3 magnitude values used to display each pixel colour in the image as displayed in Figure 21.

IMAGE PRE-PROCESSING

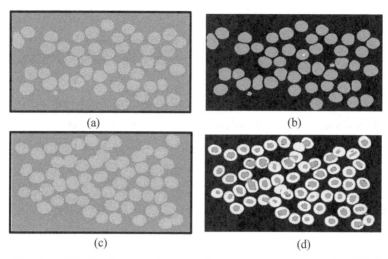

Figure 21. (a) and (c) The foreground extracted sample images converted to YC_bC_r colour space; (b) and (d) represent the recoloured down sampled image [represented with 4 colours in RGB colour space (The fourth is black)]

Owing to the variance in colour distribution among the different blood smear images, a majority based alignment rule was implemented to automatically define the region of interest cluster, i.e. the cluster inclusive of the White Blood Corpuscles(mostly nucleus material) and Malaria parasite (if any). The majority based alignment rule implemented follows the principle of identification of abnormality based on the deviation of colour code combination in comparison to 90% of clusters in a given image. It is based on the following assumptions, namely,

1) The red blood cell cluster/s (defined as a connected component/s in the masked image) will by far outnumber the number of white blood cell particles, Malaria parasite for any given image in the dataset.
2) The number of normal red blood cell will be significantly greater than the number of infected red blood cells in an image.

Apart from medical conformation of the assumptions, statistical tests were also conducted to study the ratio of red blood cell and white blood cell particles, Malaria parasite for the images in the concerned datasets. Tests were also conducted to document the ratio of normal red blood cells to infected red blood cell particles in a given image in order to verify the validity of the assumptions made. With reference to the dataset developed for the study, the median ratio of white blood cell to red blood cell cells in images containing white blood cell along with red blood cell was found to be 1.6 : 44.2 – 2 : 44 (after rounding (since average value across images results in decimal/floating point numerical estimates)). Median of the ratio of infected cells to normal red blood cells in infected images alone was found to be 4.6 : 46.6 – 5 : 47 (after rounding (since average value across images results in decimal/floating point numerical estimates))

The majority based alignment rule helps to mark out connected components containing Malaria parasite infection and white blood cell particles from red blood cells. The algorithm for the majority based alignment rule deals with the colour analysis of each connected component. Each 3-colour based connected component is divided into 3 segments (Figure 22), namely,

1) The central region
2) The intermediate region, i.e. the region lying in between the central and
3) The edge region

The colour value for each pixel is compared based on the five conditions. They are

a) If the colour of the edge and the colour of the central region for a connected component are same while the colour of the intermediate region is different and the total number of unique colours is greater than 2
b) If the colour of the central, intermediate and edge region is same and the number of unique colours for a connected component is greater than or equal to 2
c) If the colour of the edge and intermediate region is the same while the colour of the centre is different and the total number of unique colours is greater than or equal to 2
d) If the colour of the centre and intermediate region is the same while the colour of the edge is particularly different and the total number of unique colours is greater than 2
e) If the colour of the centre, edge and the intermediate region are distinctly different and the total number of unique colours is greater than 2

Based on the aforementioned conditions, the algorithm considers a particular pixel Cluster as suspect if 90% of the other connected components in the same image do not have the same number of unique colour composition like the pixel Cluster under consideration.

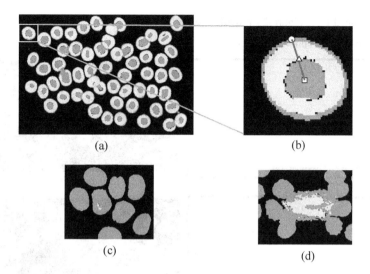

Figure 22. (a) The recoloured sample image where all red blood corpuscles are normal (b) The enlarged view of a normal cell component in the image representing the point selection process adopted for analysing each connected component (c) The recoloured image part from another image in the dataset representing presence of Malaria in an erythrocyte.[The green region represents the infection] (d) Representation of a white and red blood cell cluster from the same

Figure 22 provides an overview of the selection procedure put in place to mark out three regions of a particular connected component. The central region particularly is sampled in terms of 5 points, the centroid, and four other points in north, south, east and west at a distance of 1 pixel from the centroid. As representation of the central region the mode value of the 5 pixels is assigned as the colour of the central region. Again, the outer edge is considered as the edge region and 5 points are sampled from the edge of the connected component. Colour based Mode of the 5 pixel points is considered as the representative edge colour for a connected component. The midpoint of the centroid and each of the 5 other points on the edge is considered as representative of the intermediate region. Mode of the 5 intermediary points is considered to determine the colour of the intermediate region which is considered to be equidistant from the centre and edge alike.

Based on the assigned colour values of the three zones of the component and the aforementioned rule, the algorithm makes a decision as to whether a particular connected component has a suspect region or not. The suspect region entails White Blood corpuscles, the Malaria parasite/s (present within and outside Red Blood Corpuscles). Based on the aforementioned rule, Figure 23 provides output of segregation of suspect regions from the normal regions in a thin blood smear image.

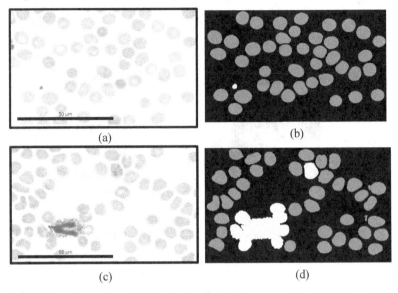

(a) (b) (c) (d)

Figure 23. (a) Input image from MaMic dataset containing normal image (b) The output image identifying only the platelet as a suspect region [Platelets eliminated based on Tukey's hinge](c) Input image containing a parasite within a red blood cell and a white blood cell (d) The output image showing the suspect region.

5.4.5. Declumping

A closer inspection of the suspect region based images (as represented in Figure 23(c)) elucidate the clustering of white blood cell with red blood cells in most images. Again, certain images in the concerned datasets were also found to contain clustering of infected and normal red blood cells which are treated as a single connected component. This shall affect Parasitaemia estimation. Herein, lies the need for de-clumping. There are 3 basic approaches that could be used for clump selection, namely,

(i) Area driven Third quartile based clump identification

Clump area >= Mean/Median_Area_of_Connected_Components x (3/2)

(ii) Tukey's Hinge based identification of Clump area

Clump area >= Mean/Median_Area_of_Connected_Components +

$2 \times \sigma(Area\_of\_Connected\_Components)$

(iii) Complete enumeration of each of the connected component.

While Tukey's upper bound for marking out clump area excludes small clumps and takes only into consideration significantly large clumps, the third quartile based threshold function identifies all of the clusters in the image.

The clusters marked out using the 3rd quartile threshold are subjected to de-clumping. For de-clumping of the clusters, the watershed algorithm was used. As a precursor to using the conventional watershed algorithm, processing of the clumped image was performed in order to automatize the process of de-clumping. The processing performed is described below.

1. For the binary cluster the distance of each point on the cluster (represented as 1) to the closest non-zero point was calculated.
2. The distance matrix (D) (of same size as the Image was calculated)
3. The distance matrix was divided into 3 by 3 overlapping blocks and the distance of each pixel in the 3 by 3 matrix from the center pixel was calculated.
4. The distance of all the pixels in the 3 by 3 matrix from the central pixel was added (and named as E).
5. The first quartile for the distance sum matrix was calculated to define a new lower bound for the distance matrix D. Values lesser than 50% of the first quartile were rejected.
6. Watershed was calculated for the distance matrix D.

Figure 24 provides a detailed stepwise output of the automated clump selection and de-clumping performed by the proposed algorithm.

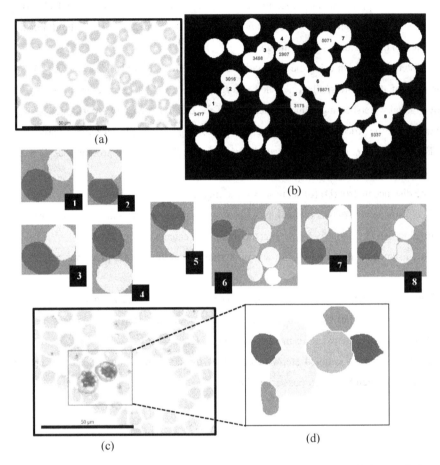

Figure 24. (a) The original image (b) The binary mask in (a) is represented along with the area of the clumped cells. The marked out clumps in the image are numbered in order to match the same with the corresponding de-clumped versions. (c) Image from the dataset consisting of a mixed cluster of RBC and white blood cell/s. (d) Image in (b) consists of only one cluster. De-clump of the mixed white and red blood cell/s.

The de-clumping algorithm is not only constricted to de-clumping connected red blood cells, the same algorithm can also be used to de-clump the cluster formed of white blood cell/s and red blood cells (Figure 24). Figure 25 shows the final de-clumped image that was reconstructed based on the de-clumping of the selected clusters.

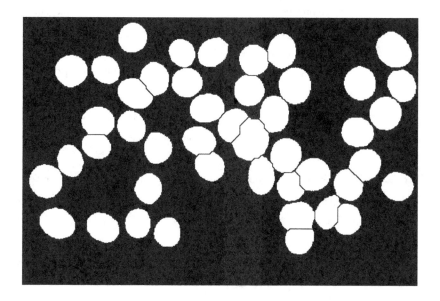

Figure 25. The reconstructed de-clumped binary image mask for the image in Figure 24a

The algorithm calculates the area of each of the components in the de-clumped image. The area value of the connected component which is greater than the Tukey's upper bound (>(median of area+ 2x standard deviation of area) value of the area of each of the connected components in the image is marked as White Blood cell while the rest are marked as red blood cells (Figure 26)

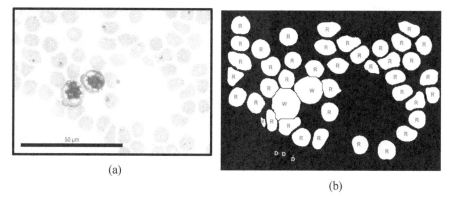

(a)

(b)

Figure 26. (a) A sample image containing infected RBC and two WBC clumped with other infected/non-infected red blood cells. (b) The binary mask of the de-clumped and reconstructed image with the white blood cells marked with 'W', the red blood corpuscles marked with 'R' and the elements to be discarded marked with D.

The problem cluster based on cluster encoding consists of the nucleus of the White Blood cell along with Malaria infection (represented in Green colour code in Figure 27c). Now the pixel positions of the problem cluster is compared with the pixel position of the mask in Figure 27b. For some images however, over segmentation was recorded for White Blood Cell Cluster. Figure 27b provides the reconstructed de-clumped binary mask obtained for the original thin blood smear image represented in Figure 27a. Again, Figure 27c is a YC_bC_r colour model based 4 colour only recoloured image. Figure 27d is a binary representation of the initial problem area cluster.

The two images, namely, the de-clumped binary mask (Figure 27b) and the binary initial problem area mask (Figure 27d) were used for pixel position matching for each of the images under consideration. The pixel position match was evaluated based on a formulated rule. The rule suggests, if an intersection in pixel position between the pixels (marked as '1') in the suspected problem cluster and the reconstructed de-clumped image is obtained and if the clump under consideration has been marked as white blood corpuscle, then the region under evaluation is removed from the suspected problem area cluster. If an intersection is coupled with a different (i.e. match with erythrocyte clump) or no match then the region is marked as problem area, Malaria parasite. Again the connected component/s that has/have been marked as normal in the suspected problem cluster, may be found to simultaneously have intersections with two or more clusters at the same time (Figure 27).

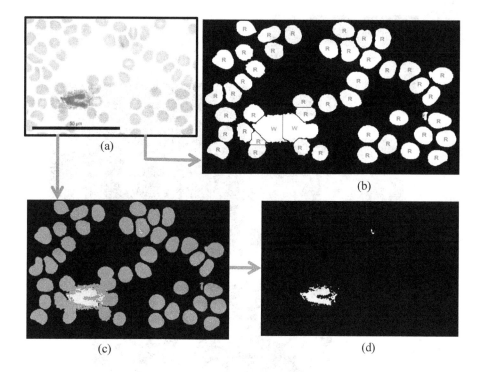

Figure 27. Representation images that were used for pixel position based matching in order to differentiate the normal (i.e. white blood cell nucleus) from the Malaria parasite in the suspected problem area cluster (a) Original digitized thin blood smear sample image ; (b) The de-clumped binary mask developed; (c) The 4 colour coded image in the YC_bC_r colour space; (d) The suspected problem cluster represented in binary

If these clusters in the de-clumped image are marked as white blood cells, then the algorithm merges these clusters into a single unit based on the intersection of the line joining the centroid of each clump to the centroid of the entire set of clumps under consideration to the clump dividing lines (Figure 28).

This approach aids towards re- clumping of single white blood cell clusters that might get de-clumped owing to pitfalls in the de-clumping algorithm

IMAGE PRE-PROCESSING

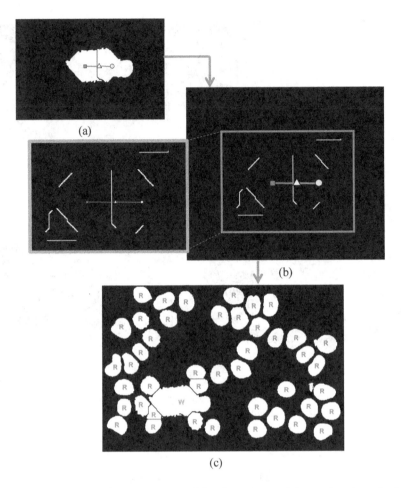

Figure 28. A brief pictorial representation of the algorithm used for re-clumping of a falsely de-clumped white blood cell. (a) The clumps (white blood cell clumps), (b) For the clump the de-clumping markers are represented in white while each of the lines (i.e. the line joining the center of each separate clump to the centroid of the clumps taken as a whole) have been represented in blue and red respectively. (c) If any of the lines intersect the de-clumping markers, the marker is removed from the reconstructed image as represented in the given figure.

5.4.6. Parasitaemia estimation

After declumping , based on area and area based percentage presence of green region, a cellular component in marked as a red blood cell component, white blood cell component or an infected RBC cluster.

IMAGE PRE-PROCESSING

The problem cluster based on cluster encoding consists of the nucleus of the White Blood cell along with Malaria infection (represented in Green colour code in Figure 29). Now the pixel positions of the problem cluster is compared with the pixel position of the mask in Figure 26b. If a match is found with red blood cell then the cell is marked out as infected. Again if a match is registered with white blood cell, then the infected cell is marked as white blood cell, hence not a problem area. If no match is found then the connected component is marked out as Malaria parasite.

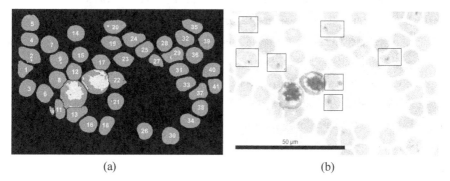

(a)   (b)

Figure 29. (a) The re-coloured image showing two separate clusters in a single frame. The image also includes erythrocyte enumeration (b) Elimination of white blood cell nucleus results in the actual marking out of the Malaria parasite infection.

Once Malaria parasite infected cells, or matured Malaria parasite is identified, Parasitaemia is evaluated again. For the image displayed in Figure 29, Parasitaemia was calculated as 6 : 41 (Malaria parasite infected cells : Total Number of Erythrocytes in the given image).

5.5. Summary

The next section particularly deals with the features that were used for parasite identification and subsequent stage, specie classification in the initial screening phase and main phase respectively.

CHAPTER 6
FEATURE SELECTION/EXTRACTION

6.1. Overview
In view of the literature at hand for Malaria parasite detection and in due consultation with a pathologist, the features used for Malaria parasite detection can broadly be classified into 2 basic types/groups, namely, Morphological features and Texture features. While Morphological features include area, eccentricity, neighbourhood cell area etc., texture features pertain to Gabor filter feature, Tamura features and the like.

Of the set of features calculated, it is again integral to identify a feature set that effectively and statistically significantly marks out Malaria parasite infection and is influential towards Malaria stage, specie classification from the red and white blood cell clusters.

Each of the two phases, in both, the initial Screening Phase and the Main Phase, Segmentation stage leads to the formation of two clusters, namely, infected RBC Cluster (Cluster 1) and WBC and probable infection cluster (Cluster 2).

6.2. Initial Screening Phase
After segregation of the red blood cell cluster from the suspect areas and WBC components, suspected RBC components were market out. Algorithmically, the red blood cells with holes were marked out based on Euler number value (Figure 30) . These binarized RBC cells were marked out as suspected RBC cell which might contain Malaria parasite at an early stage.

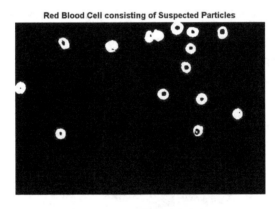

Figure 30. Red Blood Cells with suspected Malarial parasite were distinctly marked out from the RBC cluster

The edges for the distinctly marked out erythrocyte cells with suspected Malaria parasite were marked out with the Sobel operator. The suspected presence of Malaria parasite within RBC was overlaid on the edge detected image (refer Figure 31).

Red Blood Cells with Components Within

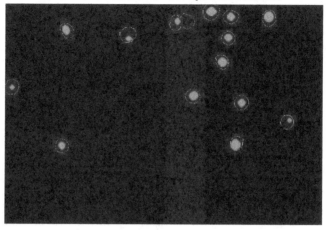

Figure 31. The suspected Malarial parasite present within RBC is marked in red. The edges of the erythrocyte cells that are deemed or suspected to contain suspected Malarial parasite are marked out

Cluster 2, consisting of WBCs, Platelets and Malaria parasites, was binarized by dynamic selection of a threshold value (Figure 32) (i.e. based on otsu's method).

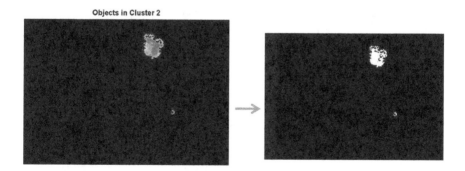

Figure 32. Cluster 2, consisting of White Blood Cells, Platelets and Malarial parasite/s (if any) was binarized by dynamic selection of threshold value

The pixels with value '1' for binarized Cluster 2 and pixels with value of '1' in Cluster 1 (i.e. the red regions in Figure 31 were marked '1') were compared. If a match is detected or an intersection is detected between the 'hole' cluster for Cluster 1 and the pixel marked '1' for

Cluster 2, Malaria parasite is said to be detected and overlaid on the blood smear image. Figure 33 represents the methodology described and followed for Malaria parasite detection present within the red blood cell (RBC).

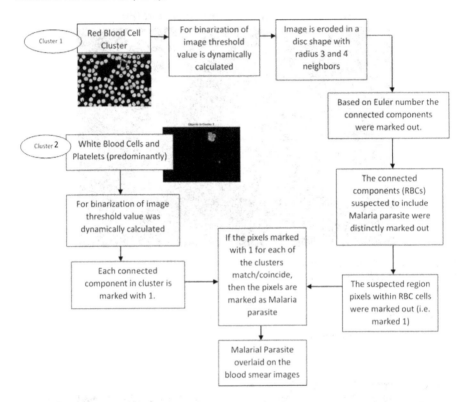

Figure 33. Flowchart representation of the algorithm followed for Malarial parasite detection within Red Blood Cells.

Figure 34, represents the overlay of the Malaria parasite present within an RBC cell on the original thin blood smear image.

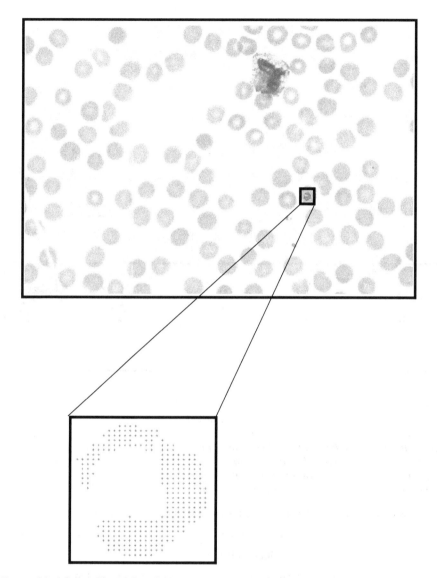

Figure 34. (above) The Malarial parasite was overlaid (with red colour) on the thin blood image (below) The detected Malarial parasite was plotted on the X-Y axis for clarity

However, neither of the methods described above can be used to trace Malaria at an advanced stage (Figure 35). Figure 35 clearly illustrate that at an advanced stage the Malaria parasite engulfs the erythrocyte and emulates the appearance of a white blood cell.

FEATURE SELECTION/EXTRACTION

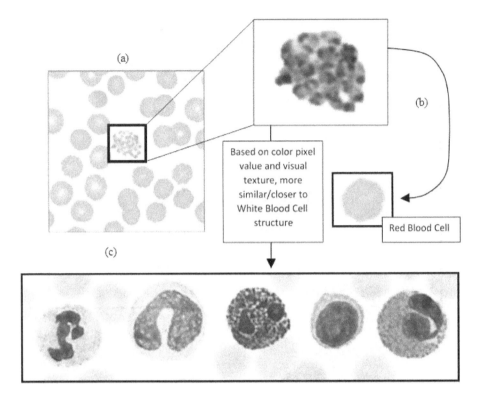

Figure 35. : (a) Thin blood smear image consisting of Malaria parasite at an advanced stage. (b) A magnified view of the Malaria parasite detected at an advanced stage. (c) In coherence with the image of the different WBCs found in blood , the image as a whole has been used as an illustration to signify that the colour pixels of the Malaria parasite at an advanced stage are more closer(in terms of Euclidean distance) to the White Blood Cells as against the Red Blood Cells. Close inspection shows that, the Malarial parasite at an advanced stage has a granulated texture similar to the texture of the White Blood Cell.

Close inspection of Figure 35 highlight that while WBC has nucleus, the Malaria parasite at an advance stage though emulates the granulated texture of WBC does not have a nucleus like structure. This distinction was algorithmically articulated.

As a result of repeated K-means clustering, three clusters were developed. The cluster consisting of WBC and platelets was further investigated for the presence/absence of Malaria parasite at an advanced stage.

Given that it is difficult to distinguish between platelets, basophil and Malarial parasite at an advanced stage, certain set of features were developed to distinguish between them. At the very onset, close proximity of erythrocyte to the particles marked out in the WBC, platelet cluster was considered. Morphological features such as dimension of all closed components such as area, perimeter, length, and breadth of each connected component was taken into account. Again, to entail a particular component as a single connected mass and not a group of small masses, the WBC/platelet cluster (i.e. Cluster 2) was dilated in a disc pattern with radius 4 and no neighbours. The values were empirically decided and are susceptible to vary for other datasets. By doing so, the connected components selected out by the algorithm resembled the individual connected components marked by an end user based on his/her cognitive ability.

Now for each of the connected components, closest neighbour detection algorithm was used to detect the presence of closest red blood cells. The number of red blood cells present within a radius of 0 to 5 pixels (calculated from the extreme most boundary point from the centroid) of a closed component in cluster 2 was calculated. Also the Euclidean distance value for the closest red blood cell was calculated. Again, the length and breadth value for each of the connected components was calculated. Given that the WBC nucleus often deviates from a regular geometrical shape, three length values and 3 breadth values were taken into account for each of the components. By dividing each closed component into 3 basic parts in a row major order the median of the breadth value for each of the 3 parts was calculated. Again dividing each closed component into 3 basic parts in column major order, the median of the length value for each of the 3 parts was calculated. The area and perimeter for each of the connected components was also calculated.

As has been previously documented (Figure 35), a basic feature that helps distinguish between a basophil and an advanced stage Malaria parasite, apart from size of particle, is the presence of nucleus within a basophil and absence of the same in an advanced stage Malaria parasite. Based on the parameter values of each of the connected components in Cluster 2, the segments representing the components were cropped out from the original blood smear image in true colour. Colour based segmentation was performed to confirm absence/presence of nucleus within a connected component (Figure 36)

FEATURE SELECTION/EXTRACTION

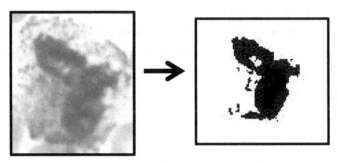

Figure 36. The segmentation of the nucleus from the White Blood Cell

To sum it all up, the features that were recorded for classification are duly marked out in Table 5

Features	Number of features
Number of red blood cells within a radius of 5 pixels from the centre of a closed component in cluster 2	1
The median length value for each of the 3 column major segments	1 x 3= 3
The median breadth value for each of the 3 row major segments	1 x 3= 3
Area of each closed component in cluster 2	1
Perimeter of each closed component in cluster 2	1
Eccentricity of each closed component in cluster 2	1
Euclidean distance of the red blood cell that is closest to the closed component in cluster 2	1
Presence/absence of nucleus within a closed component	1
Total	**12**

Table 5. The exhaustive list of the features that were calculated for each of closed components present within Cluster 2

FEATURE SELECTION/EXTRACTION

It might be effective to investigate how the aforementioned features individually contribute towards the prediction of whether a closed/connected component in Cluster 2 is a Malaria parasite or not. The feature vectors are standardized and normalized, so that all feature vectors lie in the range 0 to 1.

The model developed at the very onset segments the blood smear image into RBC and the non-RBC blood cells. While cluster 1 consists of erythrocytes, cluster 2 was found to consist of WBC, platelets and Malaria parasite (if any).

If a closed component that is part of cluster 2 exists within any of the red blood cells in cluster 1, then Malaria parasite is detected. If no such component within red blood cells can be traced, then features (Table 5) are calculated for all of the closed components in cluster 2. Based on these features a model was developed to predict whether Malaria parasite exists within the thin blood smear image.

From a range of 12 features, a feature subset of size 10 (Table 6) was marked out using Conditional Mutual Information Maximization algorithm [164].

Features	Number of features
Number of red blood cells within a radius of 5 pixels from the centre of a closed component in cluster 2	1
The median length value for the 2nd column major segments	1
The median breadth value for each of the 3 row major segments	1 x 3 = 3
Area of each closed component in cluster 2	1
Perimeter of each closed component in cluster 2	1
Eccentricity of each closed component in cluster 2	1
Euclidean distance of the red blood cell that is closest to the closed component in cluster 2	1
Presence/absence of nucleus within a closed component	1
Total	**10**

Table 6. List of selected features for classification

6.3. Main Phase

The segmentation algorithm results in the formation of two clusters, namely, the infected RBC cluster (consisting of outlier normal RBC, infected RBCs incl. artefacts) and the WBC cluster (consisting broadly of normal WBC and infection).

Based on the infection identification algorithms devised, an intersection of the output of the colour value based rule methodology (Section 5.4.4) and Lab colour based unsupervised clustering algorithm (Section 5.3), the infected RBC are identified and thereby segregated from normal RBC cells. Parts of the RBC cell (or pixels) that were marked as infection by only one of the algorithm are treated as probable infection. The intersection parts were marked as infection regions. So in particular, the RBC cell cluster was further subdivided into 2 sub-clusters, cluster of the normal RBC cells and the infected RBC cells. The probably infected cells were returned to the cluster of normal RBC cells.

The WBC cluster formed as a part of the aforementioned segmentation process consists of normal WBC cell, infection, artefacts and certain RBC cell outliers that are particularly bigger in size (based on Tukey's Hinge) than RBC and have taken a stain colour similar to the WBC cell cluster. Features were calculated for each connected component in the predominantly White Blood cell cluster. The features used to segregate a White Blood Cell from Infection and an outlier Red Blood cell can be divided into 2 broad groups: namely features from pathologist perspective and features used by other computer science researchers for identification of Malaria parasite infection. Table 7 represents the feature list that was calculated for segregation of WBC from other connected components in the cluster under consideration. The features were normalized with mean zero ('0') and standard deviation one ('1'). 3-NN classification was performed based on the feature set.

FEATURE SELECTION/EXTRACTION

Features (considering Pathologist perspective)		
Feature Name	**Number of Features**	**Time Complexity Derivation**
Nuclear Mass Presence	Number of color coded nuclear mass in a colored component ; Present/absent feature (2 features)	Search image/subimage for a particular colour shade. Given the image is in 2D, complexity is $O(n^2)$
Difference of area from Median Area of complete uninfected RBC in a radius of 2 x major axis length of the component area under consideration	Numerical feature, a negative value indicates the size of the area under consideration is smaller than the area of the average uninfected RBC in the given area and likewise for positive value (1 feature)	Identification of 8-connected components in an image – $O(n^2)$ Calculation of area of z connected component – $O(zm^2)[m \ll n]$ Calculation of median area= $O(1)$ Calculation and decision making= $O(1)$ **Total=$O(n^2)$**
Proportion of marked out infection to normal cell area(ratio of color pixel values)	Numerical float value(1 feature)	$O(1)$
Features from other authors		
Texture features –Tamura Features	3 features for each grayscale cell component	Gradient calculation – $O(n^2)$ Histogram & Peak Calculation- $O(n)$ **Total=$O(n^2)$**
Texture features – GLCM	88 features for each grayscale cell component	Generation of GLCM Matrix- $O(n^2)$ Normalization=$O(1)$ Mean, Variance, Correlation Calculation – $O(n^2)$ Other feature calculation(eg. Entropy) –$O(1)$ **Total = $O(n^2)$**
Texture features – Gabor Features	Not used	$O(n^2)$
Total Features	**94 features**	$O(n^2 + n^2 + c+ n^2 + n^2)$ = $O(n^2)$

Table 7. Feature List used for segregation of WBC Cluster from other connected components

So for species and stage classification two images have been used, infected WBC cluster and the infected RBC cluster. For the RBC infected cluster semi-supervised Lab colour based clustering was performed for pronounced ring identification within the RBC cell. The algorithm used for ring identification is particularly named as semi supervised as based on the

RGB value of the part of the ring recovered in Cluster 2 (i.e. the Cluster consisting of WBC, suspected Malaria infection etc.) region growing was performed to check the pixels in the nearby area of the suspected infection within the RBC cell that have a similar colour value in the main RGB image. Figure 37 provides a flowchart based representation of the algorithm that was developed for ring detection.

FEATURE SELECTION/EXTRACTION

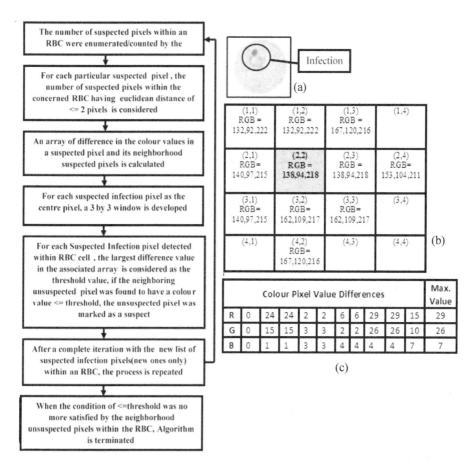

Figure 37: Flowchart of the semi-supervised algorithm for infection ring identification within RBC component. (a) A single RBC cell with P. Vivax infection in the initial stage (the infection in the cell is highlighted by way of round mark), (b) Pixels around a particular infection pixel(represented by co-ordinate (2,2) and marked in red) having euclidean distance of ≤ 2 pixel in terms of pixel X-Y co-ordinate in the image. The matrix also provides the structure of the 3 by 3 window drawn up with a particular pixel being considered to be at the centre of the window(in this case pixel having co-ordinate 2 by 2)(c) The Red(R), Green(G) and Blue(B) colour distance value of each pixel having an euclidean distance of ≤ 2 pixel from the central pixel colour value(marked in red).

Table 8 represents the features associated with pathologist perspective and other texture features that have been used for specie and stage classification of RBC cluster based infection. Features as recorded in Table 7 have been used for classification of specie and stage

FEATURE SELECTION/EXTRACTION

classification for WBC cluster. The groups in case of White Blood Cell infected cluster are *Vivax* Schizont and *Vivax* Gametocyte (Male, Female) [Two class problem]. For the RBC infected cluster the stages and specie groups are *Vivax* Ring, *Vivax* Trophozoite, *Falciparum* Trophozoite, *Falciparum* Ring, *Falciparum* Schizont, *Falciparum* Gametocyte (Male, Female).

Features (considering Pathologist perspective)		
Feature Name	**Number of Features**	**Time Complexity Derivation**
Infection Morphology	Area, Perimeter, Eccentricity (3 features)	$O(n^2)$
Cell Morphology	Area, Perimeter, Eccentricity (3 features)	$O(n^2)$
Number of infection instances in a cell	1 feature	$O(n)$
Difference of area from Median Area of complete uninfected RBC in a radius of 2 x major axis length of the component area under consideration	Numerical feature, a negative value indicates the size of the area under consideration is smaller than the area of the average uninfected RBC in the given area and likewise for positive value (1 feature)	Identification of 8-connected components in an image – $O(n^2)$ Calculation of area of z connected component – $O(zm^2)[m<<n]$ Calculation of median area= $O(1)$ Calculation and decision making=$O(1)$ **Total=$O(n^2)$**
Proportion of marked out infection to normal cell area(ratio of colour pixel values)	Numerical float value(1 feature)	$O(1)$
Features from other authors		
Texture features –Tamura Features	3 features for each grayscale cell component	Gradient calculation – $O(n^2)$ Histogram & Peak Calculation- $O(n)$ **Total=$O(n^2)$**
Texture features – GLCM	88 features for each grayscale cell component	Generation of GLCM Matrix- $O(n^2)$ Normalization=$O(1)$ Mean, Variance, Correlation Calculation – $O(n^2)$

		Other feature calculation(eg. Entropy) –O(1)
		Total = O(n²)
Total Features	**100 features**	$O(n^2 + n^2 + n + n^2 + c + n^2 + n^2) = O(n^2)$

Table 8. Feature List used for Specie and Stage Classification from RBC Cluster

For either of the two infection clusters, the feature space dimensionality is reduced to a set of 50 features using Conditional Mutual Information Maximization Algorithm [164].

Time complexity was calculated for feature extraction and subsequent multivariate filter feature set selection. The total time complexity for feature set extraction for RBC and WBC cluster was calculated as $O(n^2)$. For multivariate filter feature set selection, CMIM Algorithm was used. Based on the pseudocode provided in Fleuret's paper CMIM algorithm [164] was encoded in matlab. The code developed was estimated to have a time complexity of $O(n^2)$.

So the total time complexity for feature extraction and subsequent selection has been estimated as $O(n^2)$.

6.4. Summary

Unlike Correlation, mutual information takes into consideration the non-linear relationship between the features and labels. A particular feature might be singularly relevant for label prediction (i.e. whether a connected component is infected or normal) but it might not be significant as part of a feature set owing to data redundancy or overlapping, hence to identify a filter feature set as a whole, conditional mutual information maximization algorithm was used both in the initial screening and main phase of model development.

The next Chapter deals with the use of the features in the initial screening and main phase for cellular component classification in terms of stage and specie.

CHAPTER 7
CLASSIFICATION

7.1. Overview

The features selected/extracted in Chapter 6 were used towards connected component segregation and thereby image classification in the initial screening and main phase of data model.

7.2. Initial Screening

The feature vectors are standardized and normalized, so that all feature vectors lie in the range 0 to 1. The model developed at the very onset segments the blood smear image into RBC and the non-RBC blood cells. While cluster 1 consists of erythrocytes, cluster 2 was found to consist of WBC, platelets and Malarial parasite (if any).

If a closed component that is part of cluster 2 exists within any of the red blood cells in cluster 1, then it was marked as Malaria parasite in the initial stage. If no such component within red blood cells can be traced, then features (Table 5) are calculated for all of the closed components in cluster 2. Based on these features a model was developed to predict whether Malaria parasite exists within the thin blood smear image or outside. This part of the algorithm helps segregate Malaria parasite at the initial stage from Malaria parasite at maturity.

Based on the features calculated for each of the thin blood smear images, a supervised system was developed to predict whether cluster 2 consists only of White Blood cells and platelets (denoted by zero) or whether it consists of Malaria parasite (denoted by one) along with White Blood cells and platelets. Once the features were normalized and standardized, leave-one-out cross-validation was performed to prevent overfitting of data. As described in Figure 38, the data model developed for analysing whether a thin blood smear image does or does not contain Malaria parasite consists of 2 basic parts- a rule based part and a supervised model. Based on the prediction of the rule based section, an autonomous decision is made by the system as to whether it wants to explore the supervised learning model or not. If Malaria parasite is detected within RBC cells, the thin blood smear image is predicted to be obtained from an individual infected with Malaria. Again, if no Malaria parasite is detected within the red blood cell cluster, the closed components in cluster 2 shall be investigated using supervised learning techniques before a final conclusion can be arrived at in terms of presence/absence of Malaria parasite in a given blood smear image.

CLASSIFICATION

All features extracted from cluster 2 were normalized and standardized. Leave-one-out cross validation was performed for the images. Each image was defined as a set of datapoints where

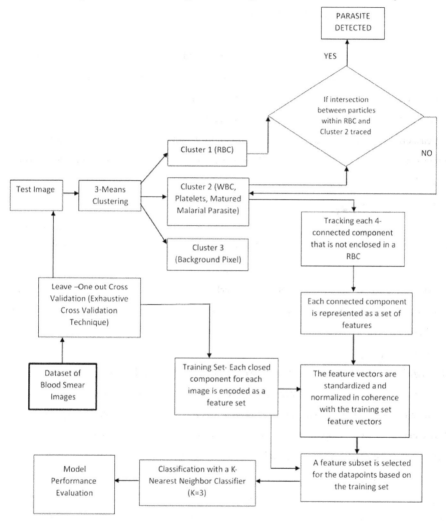

Figure 38. Data model developed for identification of Malarial Parasite within Thin Blood Smear Image

each datapoint particularly represented a closed component for cluster 2. Only the closed components that were not enclosed by a red blood cell were taken into account. A K-NN classifier was used to predict whether a closed component was a Malaria parasite at a matured state (denoted by 1) or not (denoted by 0). So each datapoint from a set of test data points for a particular image was classified based on 3 of its closest neighbours from the training set, selected on the basis of Euclidean distance (i.e. 3-NN classifier).

Each image in the dataset accounted for approximately 4 closed components in cluster 2, while some of the images recorded 6 closed components others recorded 2 and 0. In total the dataset consisted of 1012 datapoints each representing a closed component in cluster 2 that wasn't encapsulated within a red blood cell or erythrocyte. Of the 1012 datapoints, 500 datapoints represented Malarial parasite at an advanced/matured state. The other 512 datapoints represented White Blood Cells and platelet fragments.

If atleast one of the datapoints from a set of datapoints which is a part of cluster 2 for a particular image was assigned label 1, then the blood smear image as a whole was considered to be infected by Malaria parasite.

7.3. Main Phase

The 50 feature strong dataset of 1410 MaMic database images, 1320 images from the acquired dataset and mixed dataset consisting of 2730 images were all subjected to 10 fold cross validation to prevent overfitting of the data model developed.

Classification for either infected cluster is performed in three different ways to test the performance among singular classifiers and also compare their performance to ensemble classifiers.

7.3.1. Singular Classifiers

The algorithms followed by different classifiers for marking out the decision boundary vary. Given that the feature set is a filter feature set, it is not wrapped around or particularly effective for any particular classifier. The same feature set was used for classification by a K-NN, Naïve Bayes classifier and Support Vector Machine classifier singularly and subsequently their predictions were amalgamated on a majority based voting scheme.

7.3.1.1. K-NN Classifier

The value of K (i.e. the number of neighbours was varied in the range 3,5,7,9 respectively. Given that across all the datasets a neighbour size of 3 gave the optimal accuracy, sensitivity and specificity, a 3-NN classifier was used for classification purpose.

7.3.1.2. Naïve Bayes Classifier

In the Naïve Bayes Classifier all the feature vectors were assumed to follow Gaussian distribution.

7.3.1.3 Support Vector Machine

Based on experimentation Gaussian kernel of SVM was used for classification. Again, the value of C was varied in the range {1, 10, 100, 1000} while parameter Gamma was tested in the range of { $\frac{1}{2^{-15}}$,.....0.....$\frac{1}{2^{15}}$ }. Extensive Grid Search was undertaken to identify the optimal C and Gamma value combination for a RBF kernel, in terms of Accuracy, Sensitivity and Specificity value. The optimal C and gamma value selected are 1 and $\frac{1}{2}$ respectively.

7.3.2. Ensemble Classifiers

Apart from the singular classifiers, ensemble classifiers were used for further enhancing the prediction accuracy of the proposed data model. Ensemble based classification was achieved using majority based voting strategy of different type of classifiers and an Adaboost Classifier. A majority based voting of the three singular classifiers (i.e. SVM, KNN, Naïve Bayes) was implemented. The performance of the same was compared to the Adaboost Classifier that is based on an ensemble of Decision Trees, each of Level 1. The number of decision trees used for the Adaboost classifier was decided based on experimentation. For the feature space defined by 50 features, the classification accuracy for the different number of decision trees was plotted and the number of trees after which the data model was found to over-fit to the training data was considered the optimal number of trees to be used for and by the model.

Figure 39 represents the number of trees vs. Accuracy plot for the mixed dataset under consideration. Based on the plot in Figure 38, the forest was pruned and thereby made to consist of only 57 decision trees. The number of decision trees was automatically decided by the algorithm based on peak accuracy value for the test dataset under consideration.

CLASSIFICATION

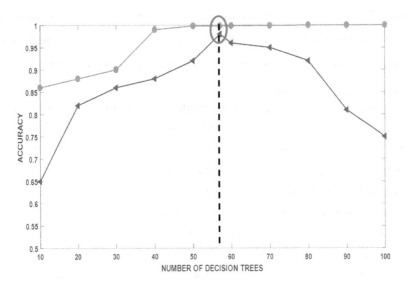

Figure 39. The Accuracy vs. Number of Decision Trees plot to identify the optimal number of Decision Trees used by the Adaboost Algorithm

The classification segment following the segmentation process for the main phase of the algorithm has been duly documented in the form of a flowchart in Figure 40

CLASSIFICATION

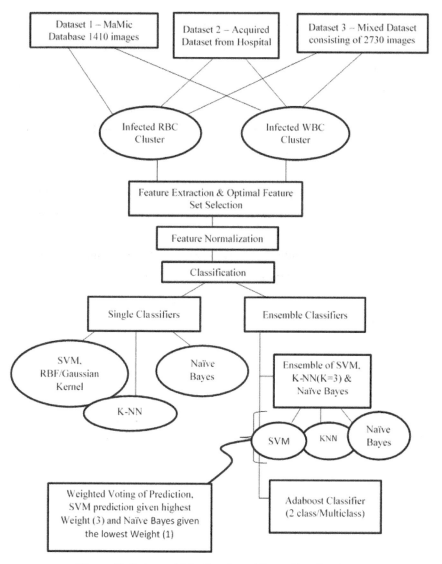

Figure 40. Data-model for Specie and Stage Classification

7.4. Summary

The use of appropriate classification system based on the features extracted is vital for the research model proposed on Malaria parasite detection and analysis. Two different approach of classification schemes was used for the respective phases (initial screening phase for detection and detailed analysis phase for detection and subsequent stage/specie classification).

CLASSIFICATION

The initial detection phase used a simple k-NN classifier that was adequate for the 10 feature based wrapper feature set as discussed in the previous chapter. However, the detailed analysis phase was more complex as it involved identification of specie and stage of infection. A filter feature set was developed for this phase and both singular and ensemble classification schemes have been implemented with the obtained feature set. Singular classification scheme used a K-NN, Naïve Bayes and Support Vector Machine classifier individually with the filter feature set at hand. Whereas, the ensemble classification scheme implemented a majority based voting of the three singular classifiers (i.e. SVM, KNN, Naïve Bayes) and an Adaboost classifier for parasite stage/specie estimation.

As evident, the Classification phase broadly has two basic phases, namely, identification of normal from infection and thereby stage, specie based classification of the infection under consideration.

Having documented the classification phase, Chapter 8 documents the evaluation results obtained for the 3 datasets under consideration for the algorithm/s developed. Performance evaluation in terms of prediction accuracy, sensitivity, specificity and time complexity is critical for the selection of certain methodology over other/s.

CHAPTER 8
RESULTS & DISCUSSION

8.1. Overview
While the results for the Initial Screening Phase has been restricted to the performance evaluation in terms of identification of infection against normal thin blood smear images, the performance results of each section of the main phase of the algorithm extending to stage/specie classificantion has been duly documented. The performance metric for both, the initial screening phase and the main phase has been compared with state of the art methods proposed and duly presented in scientific literature by other researchers.

8.2. Initial Screening Phase
The initial screening phase deals with the screening of digitized thin blood smear images of 40X magnification. The Initial Screening phase can broadly be divided into two subsections, namely, 'Discussion pertaining to Performance Metrics' & Comparative Study.

8.2.1. Performance Metrics and Discussion
In the Screening Phase, it is interesting to note that the prediction accuracy of the model degrades when all the features (Table 5) were taken into account by the classifier for making prediction. Such degradation in prediction accuracy when all features are taken into account might be due to feature redundancy. That is to say, features that bring in similar information to the classifier often hinder the performance of the classifier. Correlation accounts for the linear relationship between two variables. A strong statistically significant Pearson's rank correlation was predicted between the class label and the number of red blood cells near or around a closed component in cluster 2 (rho=0.758, p=0.000<.01). Pearson's rank correlation was used as the variables under consideration were found to deviate from normal distribution. Again, when the median of the length values for each of the 3 row major segments for each closed components was considered, a weak statistically insignificant correlation with the class label (rho =0.05, p=.534 >.01) was recorded. The correlation of the area and perimeter values of the closed components with the class labels was found to be statistically significant (rho=0.866, p=0.001<0.01). Additionally, the correlation of eccentricity value of each closed component to the class label was found to be insignificant (rho=0.25, p=0.324>0.01), while the distance of the closest RBC to the closed component was found to be significant (rho=0.665, p=0.000 <0.01) towards parasite class prediction. The presence/absence of nucleus present within a closed component in cluster 2 was found to have strong correlation with class label and was

statistically significant (rho=0.925, p=0.000>0.01). Again, correlation of class label and length, area, perimeter, breadth of nucleus present within a particular closed structure was statistically significant [rho(length)=0.730 (p=0.001<0.01), rho(area)=0.643 (p=0.001<0.01), rho(perimeter)=0.795 (p=0.000<0.01), rho(breadth)=0.712 (p=0.000<0.01)].

Conditional mutual information maximization algorithm [164] was used to find a filter feature set that bring in the maximum information about the concerned class label. This also encompasses or takes into account the non-linear relationship between a feature vector and the class label.

A minimum error rate was achieved when a subset of 10 features (Table 6) were used for prediction of the class label for a test blood smear image. Again, an exhaustive search over the feature space confirmed that the set of 10 features (i.e. formed by excluding the median length value for the 1st and 3rd half of the WBC) as a wrapper feature set provided the minimum error rate of classification for a 3-NN classifier. Table 9 represents the Accuracy, Sensitivity and Specificity values when the set of 10 features are used for identification of Malaria parasite for a thin blood smear image by varying the number of neighbours (K=1,3,5) for a test datapoint.

Number of Neighbours	Accuracy	Sensitivity	Specificity
K=1	0.91898	0.92	0.91797
K=3	**0.97826**	**0.98**	**0.97656**
K=5	0.94169	0.95	0.93359

Table 9. The Accuracy, Sensitivity and Specificity values when the Neighbours used for classification of a test datapoint were varied between 1, 3 and 5

A rule based methodology was developed for the same set of images. A rule set replacing 3 – Means clustering was developed and appended to the algorithm developed. A particular set of colour intensity values in the HSV colour model were assigned to cluster 1, 2 and 3 respectively by virtue of a set of defined rules. The accuracy, sensitivity and specificity of assignment of pixels by the rule based system have been reported against the accuracy, sensitivity and specificity values for the unsupervised 3-means clustering system (Table 10). The threshold range for the different feature vectors were empirically worked on. It was difficult to narrow in on a suitable threshold range for feature vectors (namely the median of the length values for the three row major segments for the closed components) that did not have a statistically significant correlation with the class labels. Again, it was tedious to tune the model to perform satisfactorily when the magnification of the concerned images was varied as it also lead to re-estimation of the threshold values. For the developed rule based system, the classification accuracy, sensitivity and specificity values are represented in Table 10.

	Accuracy	Sensitivity	Specificity
RULE BASED METHOD			
Rule based pixel clustering	0.62	0.63	0.66
Overall Classification of Malarial parasite	0.741107	0.706	0.77539
MACHINE LEARNING BASED METHOD			
3-Means clustering	0.9895	0.9925	0.9832
Overall classification of Malarial parasite	0.7260	0.6895	0.7123

Table 10. Comparison of performance metrics between Rule base and ML based algorithms

The method proposed in this work is a hybrid or a generous optimized mixture of rule based and machine learning based approach. While a singularly rule based system has its drawbacks it might be helpful to analyse the drawbacks of using a machine learning based methodology singularly. To test the effectiveness of making the algorithm particularly machine learning based in entirety certain changes were incorporated in the algorithm developed. At the onset, the rule based identification of Malarial parasite within an RBC by matching the closed component present within RBC to the closed components in cluster 2 was removed. The change that has been introduced into the algorithm has been highlighted in Figure 41. All 4-connected components in Cluster 2 were taken into consideration and each connected component was treated as a datapoint that was described by the set of features marked out in Table 5. Based on the feature values and leave one out cross-validation, a K-NN classifier was used to classify each of the closed components as a Malaria or non-Malaria parasite. Similar to the proposed algorithm, the value of K was varied in the range 1, 3 and 5 respectively. The machine learning based methodology often leads to misidentification of Malaria parasite present within red blood corpuscle at the initial stage of infection. The Malarial parasite prediction accuracy, sensitivity and specificity value for the machine learning based system has been documented in Table 10. The aim of the algorithm developed was to identify/predict the presence of Malaria parasite in a given thin blood smear image at the initial, advanced or at a matured stage when the parasite engulfs the RBC. Table 9 and 10 highlight the efficiency of the hybrid algorithm for Malaria parasite prediction as opposed to the prediction accuracy of a rule based or machine learning based system.

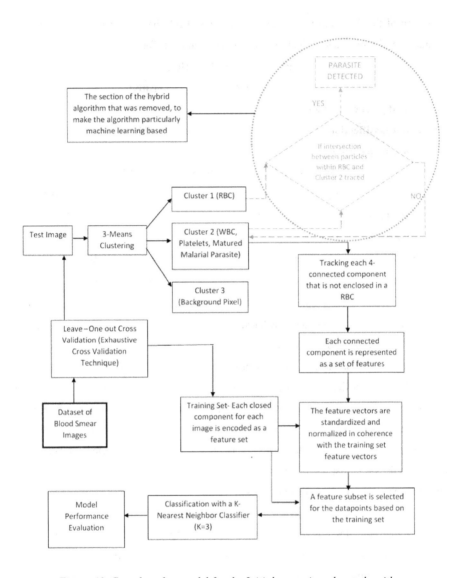

Figure 41. Complete datamodel for the Initial screening phase algorithm

8.2.2. Comparative Study

In context of the initial screening phase, two way comparative study was conducted. Table 11 documents the reported performance metric of other algorithms developed for Malaria parasite detection along with the proposed initial screening phase.

RESULTS & DISCUSSION

Algorithm	Sensitivity	Specificity	Dataset used
Boray Tek et. al. [89]	74	98	9 images
Makkapati et. al. [99]	83	98	55 images
Somasekar et. al. [108]	94.87	97.3	76 images
Diaz et. al. [112]	94	99.7	100 images
Ghate et. al. [134] (Method 1)	81.39	86.49	80 images
Ghate et. al. [134](Method 2)	72.93	75.76	80 images
Suryawanshi et. al. [150]	93.33	93.33	30 images
Mashor et. al. [148]	92.14	99.79	100 images
Chayadevi et. al. [152]	95.2092	96.06758	476 images
Nag et.al. [Proposed Hybrid Approach]**	98.4	97.6	250 images from Database

Table 11. Comparative Study of the overall performance of the proposed Screening Algorithm with other research works

A closer inspection of Table 11 reveals that the algorithms proposed by Boray Tek et. al.[89], Makkapati et. al.[99], Mashor et. al.[148] are reported to perform better than the proposed algorithm in terms of the Specificity value. While Specificity only identifies the efficiency of the system in marking out the True Negatives, Sensitivity estimates the performance of the algorithm in identifying True Poitives. So in terms of reported Sensitivity & Specificity Values, the performance of the proposed algorithm performs better than the other algorithms in Malaria parasite detection. However, the metrics are particularly incomparable because each of them uses different dataset. Hence, certain algorithms developed by other researchers were programmatically coded and used on the initial screening database to obtain the performance metric on the database at hand (Table 12). This was done to critically analyze the performance of other algorithms developed against the proposed screening algorithm in context of the 250 image strong database.

Algorithm	Sensitivity	Specificity	Dataset used
Somasekar et. al. [108]	90.4	93.6	250 images from Database
Diaz et. al. [112]	93.6	88	
Suryawanshi et. al. [150]	75.2	92.0	
Mashor et. al. [148]	74.4	77.6	
Chayadevi et. al. [152]	88.8	85.6	
Nag et.al. [Proposed Hybrid Approach]	98.4	97.6	

Table 12. Comparative Study of the overall performance of the proposed Screening Algorithm with other proposed methods by Authors using 250 Images of dataset used in the thesis.

It stands relevant to mention that the consulting pathologist provided the ground truth and selected a subset of 250 images from the MaMic dataset for evaluation of the initial screening algorithm. As per the doctor, the selected subset has adequate representation of all classes of infection and all features necessary for the screening algorithm to provide good prediction result.

8.3. Main Phase

The main phase though represented as dependent on the Initial Screening Phase, yet it is inclusive of all sections, namely, parasite detection, parasitaemia estimation and subsequent stage/specie classification. Hence the main phase can also function as an independent, reliable system. However, it has been serially connected to the Initial Screening Phase to enhance system efficiency in terms of time complexity from the perspective of a computer scientist. From clinical perspective the main phase is considered more reliable owing to the use of the accepted standard of 100X magnification of digitized blood smear images.

8.3.1. Performance Metrics and Discussion

In accordance with the proposed methodology, the algorithm for the main phase can particularly be classified into three essential blocks namely, image pre-processing along with ROI extraction, De-clumping of cluster with red blood cell enumeration and Malaria parasite detection (i.e. Segmentation) with subsequent stage/specie classification.

For the first section, the foreground identification accuracy achieved by 3-Means Clustering was found to be 92.19 % across all images in the database against the modified Zack's thresholding [127] which recorded only a value of 63.75 % for the dataset at hand. Again, these values were obtained on images for which illumination was not corrected (Table 13) in order to enable a single algorithm to be used on the two datasets under consideration without any change.

Method used for Background separation(Without illumination correction of the images)	Accuracy achieved
3-Means clustering	92.19%
Modified Zack's thresholding	63.75%
Algorithm Performance after image illumination correction	
3-Means clustering	98.96%
Modified Zack's thresholding	63.75%

Table 13. Comparative account of the accuracy achieved by the two methods used for image background separation before and after illumination correction (which was separately implemented for the two datasets at hand)

On illumination correction, the 3-NN classifier recorded an accuracy value of 98.96% as opposed to the performance of Zack's thresholding algorithm [127] that did not improve (Table 13). Thereby after illumination correction and subsequent scale and other artefact removal, 3-Means clustering algorithm was used to separate out the foreground from the background. Hence as opposed to the Foreground segregation technique proposed/suggested by Diaz et.al. [112], Linder et al. [165], unsupervised 3-Means clustering algorithm was used to mark out the foreground against the background. The values were compared against a manually marked out foreground pixel set marked as ground truth for each of the respective images in the dataset.

The second integral section can further be divided into two serially dependent blocks, namely, de-clumping and red blood cell enumeration. Deviating from the template based de-clumping technique proposed by Diaz et al [112], Kumar et al. [117] general purpose de-clumping method followed by Sio et al. [104], a modified watershed based methodology was used for automatic de-clumping of red blood cell clusters as also mixed red and white blood cell clusters. While the effectiveness of the concavity rule or the de-clumping method proposed by Devi [158] was found to be limited for the dataset at hand (Figure 28c). For the proposed algorithm, an image based dynamic threshold value for automatic selection of clumps has been suggested (Table 14). De-clumping of both, red blood cell clusters and mixed red and white blood cell clusters has been dealt with in this work.

RESULTS & DISCUSSION

Statistical Metric Used/Proposed	Accuracy	Sensitivity	Specificity
Use of third quartile as a threshold	100 %	1	1
Use of Tukey's upper hinge as a threshold	97.82%	0.3829	0.9982

Table 14. Performance evaluation for clump identification using two separate thresholds

As evident from Table 14, the third quartile of area distribution was used as the metric of choice for the database under consideration. The proposed de-clumping algorithm achieved Sensitivity and Specificity value of 0.9333 and 0.9807 respectively. Again, based on the nucleus structure, over-segmented white blood cells were re-clumped.

After de-clumping, red blood cells were segregated from white using a threshold value. Two threshold values were tested to segregate red from white blood cells, namely, Tukey's upper hinge value for the de-clumped area distribution and (mean+ 3σ). Table 15 represents the performance metric for red blood cell segregation from white blood cells. Again, based on the nucleus structure, over-segmented white blood cells were re-clumped.

Statistical Metric Used/Proposed	Accuracy	Sensitivity	Specificity
Use of Tukey's upper hinge as a threshold	96.6%	0.9770	0.83333
Use of (mean+ 3σ) as threshold	97.73%	0.9944	0.7593

Table 15. Red Blood Cell segregation from White Blood Cell in the digitized thin blood smear image dataset – performance metric using two different threshold (for 1st iteration).

Two stage/ iteration based de-clumping was performed to improve the segregation of red and white blood cells. The segregation method is referred to as a 2 stage/iteration based de-clumping method because at the 1st. iteration, the clusters are identified using the third quartile value of area as a threshold. Once de-clumped, Tukey's upper hinge of the area variable was

RESULTS & DISCUSSION

calculated based on the de-clumped image was used as a threshold to segregate red blood cells from white blood cells. Based on the green colour nucleus material re-clumping was performed in case of White blood cell components. Table 16 represents the performance metric for red blood cell segregation from white blood cells for each of the databases under consideration.

Dataset	Accuracy	Sensitivity	Specificity
MaMic Database	0.9925	0.9875	0.9942
Acquired Database	0.9865	0.9944	0.9793

Table 16. Red Blood Cell segregation from White Blood Cell in the digitized thin blood smear for the two datasets under consideration

While the first section forms the basis of the algorithm as a whole, the second block is precursor to Parasitaemia estimation that is worked on in the third block of the research design. The third block is aimed to identify presence of Malaria parasite in a thin blood smear image. The RGB image was represented in the YC_bC_r space in order to down sample the large number of colours used to represent a single connected component. Based on the YC_bC_r values the image was recoloured to four basic colours, namely, red, green, blue and black where black represents the background of the image. Malaria parasite, within or outside red blood cell and white blood cell nucleus were both represented in green colour code for all images. The pixel position values in the green clusters were compared to the positions in binary mask developed. If a match exists, then the green connected component is a part of the component in question (white blood cell, red blood cell). A white blood cell is identified based on the presence of nucleus (represented in green) and area value is marked as normal, the other problem area/s present within a red blood cell or as standalone are retained for further investigation. The performance metric of the proposed system for Malaria parasite detection has been duly recorded in Table 17. The values in Table 17 pertain to the mixed Dataset under consideration.

Algorithm Evaluation Parameter	Value obtained
Accuracy	98.11%
Sensitivity(True Positive Rate)	0.9645
Specificity(1- False Positive Rate)	1
Area under Curve (AUC)	0.9583

Table 17. Detection Algorithm performance evaluation metric

Figure 42 represents the Receiver Operating Characteristic Curve obtained for TPR values obtained for 30 threshold points.

RESULTS & DISCUSSION

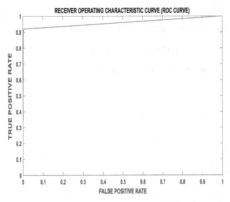

Figure 42. The ROC curve for the estimation of the algorithm performance

For Parasitaemia estimation, the algorithm identifies a white blood cell based on the presence of nucleus (represented in green) and area value along with other texture features as highlighted in Table 7. On identification a WBC is marked as normal, while the other problem area/s present within a red blood cell or as standalone are retained for further investigation (Figure 43).

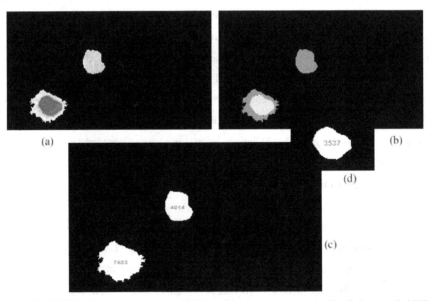

Figure 43. A WBC cluster containing a WBC and a vivax gametocyte. (b) Colour coded WBC nucleus and infection, the green represents the WBC nucleus. (c) represents the area of the gametocyte and the WBC, (d) area of the nucleus that has been used as a feature for segregating WBC from Malaria infection in WBC Cluster

RESULTS & DISCUSSION

As per the proposed model, the performance of the hybrid algorithm towards detection of infected RBC at cellular level was recorded (Accuracy 0.9962, Sensitivity 0.9963, and Specificity-0.9949). In case of the WBC Cluster, a 3 –NN classifier was used to separate out white blood cell, infection, artefact (i.e. clustered platelets) and outlier red blood cell. Figure 44 and 45 infections traced in the infected WBC and RBC cluster along with certain new morphological features used. In Figure 44b, the radius used to Table 18 represents an excerpt of the 88 feature values that were statistically significant towards differentiating a *Vivax* Gametocyte from a *Vivax* Schizont in the MaMic Database.

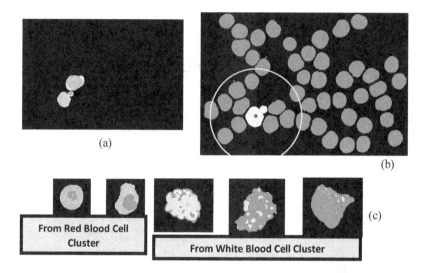

Figure 44. (a) WBC cluster consisting of RBC joined with Platelet Artefact (b) the radius feature used for estimation whether the RBC should be considered as normal or bigger in size (infection/ outlier) (c) Different stages of infection P.vivax detected by the main phase of the proposed algorithm

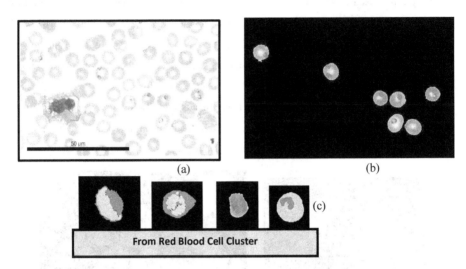

Figure 45.(a) Original Image from MaMic database, (b) Infected RBC Cluster with Ring Identification (based on algorithm represented in Figure 37) (c) P. falciparum infection at different stages as detected by the main phase of the proposed algorithm

Feature	Prerequisite	Test Performed
Autocorrelation (0°)	Shapiro Wilk Test W(69)=0.892, p=.001<0.05	Mann Whitney U Test U=0, p=.000<.05 [significant difference]
Autocorrelation (45°)	W(69)=.887,p=.001<.05	U=0, p=.000<.05 [significant difference]
Autocorrelation (90°)	W(69)=.861,p=.000<.05	U=0, p=.000<.05 [significant difference]
Autocorrelation (135°)	W(69)=.886,p=.001<.05	U=0, p=.000<.05 [significant difference]
Cluster Shade(0°)	W(69)= .735,p=.000<.05	U=0, p=.000<.05 [significant difference]
Cluster Shade(45°)	W(69)=.728,p=.000<.05	U=0, p=.000<.05 [significant difference]
Cluster Shade(90°)	W(69)=.719,p=.000<.05	U=0, p=.000<.05 [significant difference]
Cluster Shade(135°)	W(69)=.731,p=.000<.05	U=0, p=.000<.05 [significant difference]

RESULTS & DISCUSSION

Dissimilarity($0°$)	W(69)=.943,p=.038<.05	U=0, p=.000<.05	[significant difference]
Dissimilarity($45°$)	W(69)=.941,p=.030<.05	U=0, p=.000<.05	[significant difference]
Dissimilarity($90°$)	W(69)=.896,p=.001<.05	U=0, p=.000<.05	[significant difference]
Dissimilarity($135°$)	W(69)=.908,p=.003<.05	U=1, p=.000<.05	[significant difference]
Energy($0°$)	W(69)=.864,p=.000<.05	U=0, p=.000<.05	[significant difference]
Energy($45°$)	W(69)=.888,p=.001<.05	U=, p=.000<.05	[significant difference]
Energy($90°$)	W(69)=.858,p=.000<.05	U=0, p=.000<.05	[significant difference]
Energy($135°$)	W(69)=.859,p=.000<.05	U=0, p=.000<.05	[significant difference]
Entropy($0°$)	W(69)=.909,p=.003<.05	U=0, p=.000<.05	[significant difference]
Entropy($45°$)	W(69)=.923,p=.008<.05	U=11, p=.000<.05	[significant difference]
Entropy($90°$)	**W(69)=.953,p=.080>.05**	Levene's Test(F)=.806(p=.375>0.05), t(67)=-9.783(p=0.000<0.05) [significant difference]
Entropy($135°$)	**W(69)=.956,p=.109>.05**	Levene's Test(F)=3.521(p=.068>0.05), t(67)=-9.626(p=0.000<0.05) [significant difference]
Sum of Squares Variance($0°$)	W(69)=.874,p=.000<.05	U=0, p=.000<.05	[significant difference]
Sum of Squares Variance($45°$)	W(69)=.879,p=.000<.05	U=0, p=.000<.05	[significant difference]
Sum of Squares Variance($90°$)	W(69)=.852,p=.000<.05	U=0, p=.000<.05	[significant difference]
Sum of Squares Variance($135°$)	W(69)=.891,p=.001<.05	U=0, p=.000<.05	[significant difference]

RESULTS & DISCUSSION

Sum Average(0°)	**W(69)=.982,p=.741>.05**	Levene's Test(F)=.525(p=.473>0.05) , t(67)=7.729(p=0.000<0.05) [significant difference]
Sum Average(45°)	**W(69)=.959,p=.139>.05**	Levene's Test(F)=.134(p=.716>0.05) , t(67)=9.726(p=0.000<0.05) [significant difference]
Sum Average(90°)	**W(69)=.974,p=.434>.05**	Levene's Test(F)=2.656(p=.111>0.05) , t(67)=9.001(p=0.000<0.05) [significant difference]
Sum Average(135°)	W(69)=.947,p=.049<.05	U=22, p=.000<.05 [significant difference]
Sum Variance(0°)	W(69)=.835,p=.000<.05	U=0, p=.000<.05 [significant difference]
Sum Variance(45°)	W(69)=.876,p=.000<.05	U=0, p=.000<.05 [significant difference]
Sum Variance(90°)	W(69)=.860,p=.000<.05	U=0, p=.000<.05 [significant difference]
Sum Variance(135°)	W(69)=.831,p=.000<.05	U=0, p=.000<.05 [significant difference]
Sum Entropy(0°)	**W(69)=.851,p=.071>.05**	Levene's Test(F)=6.861(p=.012<0.05) , t(59.563)=-6.794(p=0.000<0.05) [significant difference]
Sum Entropy(45°)	**W(69)=.861,p=.159>.05**	Levene's Test(F)=0.355(p=.555>0.05) , t(67)=-6.627(p=0.000<0.05) [significant difference]
Sum Entropy(90°)	W(69)=.917,p=.005<.05	U=16, p=.000<.05 [significant difference]
Sum Entropy(135°)	**W(69)=.980,p=.653>.05**	Levene's Test(F)=0.070(p=.793>0.05) , t(67)=-6.580(p=0.000<0.05) [significant difference]
Difference Entropy(0°)	W(69)=.899,p=.001<.05	U=0, p=.000<.05 [significant difference]

RESULTS & DISCUSSION

Difference Entropy(45°)	W(69)=.926,p=.010<.05	U=0, p=.000<.05	[significant difference]
Difference Entropy(90°)	W(69)=.896,p=.001<.05	U=2, p=.000<.05	[significant difference]
Difference Entropy(135°)	W(69)=.923,p=.008<.05	U=0, p=.000<.05	[significant difference]

Table 18. *The values of the GLCM features that are significant towards distinction between vivax Schizont vivax Gametocyte from MaMic dataset*

The evaluation of the algorithm for infection detection within RBC cell and for segregation of WBC from infection, artefact and outlier RBC using a One vs. all strategy for each of the three databases under consideration has been represented in Table 19. The detailed stage/specie based classification result has been duly represented in Section 8.3.1.1.

Database	Accuracy	Sensitivity	Specificity
RBC Cluster			
MaMic Database	0.9946	0.9963	0.9949
Acquired Database	0.9768	0.9742	0.9773
Mixed Database	0.9860	0.9869	0.9846
WBC Cluster			
MaMic Database	0.9623	0.9609	0.9726
Acquired Database	0.9678	0.9497	0.9705
Mixed Database	0.9649	0.9621	0.9637
Infection in WBC Cluster			
MaMic Database	0.9705	0.9714	0.9710
Acquired Database	0.9565	0.9807	0.9693
Mixed Database	0.9625	0.9770	0.9700

Table 19. *Performance statistics for identification of infected RBC from normal RBC Cells and for identification of infected WBC cell cluster using One Vs All Strategy*

The final feature set of size 50 is selected using Conditional Mutual Information Maximization Algorithm. Table 20 represents the final average classification accuracy across all infection classes with single and ensemble classifiers respectively at image level.

RESULTS & DISCUSSION

Classifier(Specie & Stage Classification)	Accuracy [For 50 feature set selected by CMIM]	Sensitivity	Specificity
K-NN(K=3)	0.9653	0.9472	0.9677
SVM(RBF Kernel) C-1, gamma- ½	0.9759	0.9531	0.9753
Naïve Bayes Classifier (Gaussian distribution assumed)	0.9245	0.9168	0.9344
Ensemble(SVM, 3-NN, Naïve Bayes)	0.9874	0.9925	0.9820
Adaboost (57 trees)	**0.9917**	**0.9948**	**0.9892**

Table 20. Performance statistics for final average classification accuracy across all infection classes with single and ensemble classifiers (at image level)

8.3.1.1. Detailed Stage/Specie based Classification (Cellular Level)

In Stage/Specie Classification, there are two paticular clusters that were separately investigated. While one cluster is of infected Red Blood cells. The other cluster investigated was of white blood cells. As evident from Figure 44 and 45, while the RBC Cluster consists of *Vivax* Ring, *Vivax* Trophozoite, *Falciparum* Trophozoite, *Falciparum* Ring, *Falciparum* Schizont, *Falciparum* Gametocyte (Male, Female) thereby leading to the formulation of a multiclass (i.e. 6 class problem). The WBC infection cluster in particular consists of P.*Vivax* Schizont and Gametocyte, thereby transforming it to a 2-class problem.

8.3.1.1.1. Red Blood Cell Cluster

Identification of infection present in the RBC Cluster includes the parasites enclosed by a Red Blood Cell. The algorithm utilizes the RBC Cluster that only includes infected Red Blood cells and classifies them into 6 groups. Figure 46 shows the representation of a confusion matrix and the performance metrics calculations. Figure 47 shows Confusion Matrices of MaMic Dataset, Figure 48, shows Confusion Matrices of Acquired Hospital Dataset and Figure 49 shows Confusion Matrices of mixed Dataset

RESULTS & DISCUSSION

Confusion Matrix

True Positive (TP)	False Positive (FP)
False Negative (FN)	True Negative (TN)

Sensitivity: TP/(TP+FN)

Specificity: TN/(TN+FP)

Accuracy: (TP+TN)/(TP+TN+FN+FP)

Figure 46. Confusion Matrix and performance calculation

Confusion Matrix

Vivax Ring	Truth	
Predicted Value	161	4
	1	628

Vivax Trophozoite	Truth	
Predicted Value	114	4
	0	673

Falciparum Ring	Truth	
Predicted Value	367	3
	6	416

Falciparum Trophozoite	Truth	
Predicted Value	38	2
	0	752

Falciparum Schizont	Truth	
Predicted Value	28	4
	0	760

Falciparum Gametocyte	Truth	
Predicted Value	74	2
	0	714

Performance Statistics

Infection Type Specie & Stage	Sensitivity	Specificity	Accuracy
Vivax Ring	0.9938	0.9936	0.9937
Vivax Trophozoite	1	0.9940	0.9949
Falciparum Ring	0.9839	0.9928	0.9886
Falciparum Trophozoite	1	0.9973	0.9974
Falciparum Schizont	1	0.9947	0.9949
Falciparum Gametocyte	1	0.9972	0.9974
Average	0.9963	0.9949	0.9946

Performance Statistic Calculation for first Confusion Matrix (Vivax Ring):
Sensitivity: 161/(161+1)
   161/162 = 0.9938
Specificity: 628/(628+4)
   628/632 = 0.9936
Accuracy: (161+628)/(161+1+628+4)
   789/794 = 0.9337

Figure 47. Performance Metrics and Confusion Matrix for each type and stage of infection as obtained from the MaMic (dataset #1)

RESULTS & DISCUSSION

Confusion Matrix

Vivax Ring	Truth	
Predicted Value	221	19
	5	869

Vivax Trophozoite	Truth	
Predicted Value	146	22
	7	927

Falciparum Ring	Truth	
Predicted Value	520	12
	12	557

Falciparum Trophozoite	Truth	
Predicted Value	53	21
	1	1027

Falciparum Schizont	Truth	
Predicted Value	37	27
	1	1036

Falciparum Gametocyte	Truth	
Predicted Value	106	25
	2	969

Performance Statistics

Infection Type Specie & Stage	Sensitivity	Specificity	Accuracy
Vivax Ring	0.9768	0.9786	0.9782
Vivax Trophozoite	0.9542	0.9768	0.9736
Falciparum Ring	0.9774	0.9789	0.9782
Falciparum Trophozoite	0.9814	0.9799	0.9800
Falciparum Schizont	0.9736	0.9746	0.9745
Falciparum Gametocyte	0.9814	0.9748	0.9754
Average	0.9742	0.9773	0.9768

Figure 48. Performance Metrics and Confusion Matrix for each type and stage of infection as obtained from the Acquired (Hospita) (dataset #2)

Confusion Matrix

Vivax Ring	Truth	
Predicted Value	337	25
	1	1497

Vivax Trophozoite	Truth	
Predicted Value	364	27
	1	1599

Falciparum Ring	Truth	
Predicted Value	899	13
	5	976

Falciparum Trophozoite	Truth	
Predicted Value	91	24
	1	1778

Falciparum Schizont	Truth	
Predicted Value	63	31
	3	1798

Falciparum Gametocyte	Truth	
Predicted Value	180	27
	2	1683

Performance Statistics

Infection Type Specie & Stage	Sensitivity	Specificity	Accuracy
Vivax Ring	0.9973	0.9835	0.9863
Vivax Trophozoite	0.9972	0.9833	0.9859
Falciparum Ring	0.9944	0.9868	0.9904
Falciparum Trophozoite	0.9891	0.9866	0.9868
Falciparum Schizont	0.9545	0.9830	0.9820
Falciparum Gametocyte	0.9890	0.9842	0.9846
Average	0.9869	0.9846	0.9860

Figure 49. Performance Metrics and Confusion Matrix for each type and stage of infection as obtained from the Mixed dataset (dataset #3)

RESULTS & DISCUSSION

8.3.1.1.2. White Blood Cell Cluster

The WBC Cluster particularly includes the White Blood Cell that is characterized by the presence of nucleus, the platelets which takes a dark stain and often forms clusters and appears as staining artefacts, Red Blood Cell Outliers that are larger than normal RBC and often takes a darker stain and parasite infected RBC. The *P. vivax* schizont and gametocyte are larger in size and completely engulfs the RBC. Owing to their size and staining characteristics they are present in this particular cluster. The algorithm initially differentiates and classify the objects present in the WBC cluster into four groups, namely, WBC, RBC Outliers, Artefacts and Parasites using one versus all strategy for the particular classifier used. After this process of segregation of parasite infected RBC, the infected objects are classified as *P. vivax* Schizont/Gametocyte as a two-class problem.

8.3.1.1.2.1. Categorizing all connected components into 4-Classes

The Confusion matrix and performace analysis of 4-Classes of the three dataset is represented in Figure 50 for MaMic Dataset, Figure 51 for Acquired Dataset and Figure 52 for the Mixed Dataset

Confusion Matrix

White Blood Cell	Truth	
Predicted Value	124	9
	3	110

Infected RBC	Truth	
Predicted Value	67	8
	2	169

RBC Outliers	Truth	
Predicted Value	36	6
	1	203

Artefact	Truth	
Predicted Value	12	7
	1	226

Performance Statistics

Type of Connected Components	Sensitivity	Specificity	Accuracy
White Blood Cell	0.9763	0.9243	0.9512
Infected RBC	0.9710	0.9548	0.9593
RBC Outliers	0.9729	0.9712	0.9715
Artefact	0.9230	0.9699	0.9674
Average	**0.9608**	**0.9551**	**0.9623**

Figure 50. Performance Metrics and Confusion Matrix for each type and stage of infection as obtained from the MaMic dataset for WBC Cluster (dataset #1)

RESULTS & DISCUSSION

Confusion Matrix

White Blood Cell	Truth	
Predicted Value	72	5
	5	136

Infected RBC	Truth	
Predicted Value	95	5
	3	115

RBC Outliers	Truth	
Predicted Value	17	4
	2	195

Artefact	Truth	
Predicted Value	22	4
	0	192

Performance Statistics

Type of Connected Components	Sensitivity	Specificity	Accuracy
White Blood Cell	0.9350	0.9645	0.9541
Infected RBC	0.9693	0.9583	0.9633
RBC Outliers	0.8947	0.9798	0.9724
Artefact	1	0.9795	0.9816
Average	**0.9497**	**0.9705**	**0.9678**

Figure 51. Performance Metrics and Confusion Matrix for each type and stage of infection as obtained from the Acquired dataset for WBC Cluster (dataset #2)

Confusion Matrix

White Blood Cell	Truth	
Predicted Value	196	12
	8	248

Infected RBC	Truth	
Predicted Value	162	13
	5	284

RBC Outliers	Truth	
Predicted Value	53	11
	3	397

Artefact	Truth	
Predicted Value	34	12
	1	417

Performance Statistics

Type of Connected Components	Sensitivity	Specificity	Accuracy
White Blood Cell	0.9607	0.9538	0.9568
Infected RBC	0.9700	0.9562	0.9612
RBC Outliers	0.9464	0.9730	0.9698
Artefact	0.9714	0.9720	0.9719
Average	**0.9621**	**0.9637**	**0.9649**

Figure 52. Performance Metrics and Confusion Matrix for each type and stage of infection as obtained from the Mixed dataset for WBC Cluster (dataset #3)

8.3.1.1.2.2. Categorizing Infected RBC into 2-Classes in WBC Cluster

The Confusion matrix and performace analysis of 2-Classes of the three dataset for RBC infection in WBC Cluster is shown in Figure 53

RESULTS & DISCUSSION

Confusion Matrix

MaMic Dataset		
Infection Schizont/Gametocyte	Truth	
Predicted Value	33	1
	1	34

Hospital Dataset		
Infection Schizont/Gametocyte	Truth	
Predicted Value	44	1
	2	51

Combined Dataset		
Infection Schizont/Gametocyte	Truth	
Predicted Value	77	2
	3	85

Performance Statistics

Infection	Sensitivity	Specificity	Accuracy
MaMic	0.9705	0.9714	0.9710
Hospital	0.9565	0.9807	0.9693
Combination	0.9625	0.9770	0.9700

Figure 53. Performance Metrics and Confusion Matrix for each type and stage of infection as obtained from the all three dataset for WBC Cluster

8.3.1.2. Image-Level Classification of Infected and Normal Images

After identification of infection in the two clusters it is logically relevant to report whether the concerned image is normal/infected. The Classification of image based on Ada-boost is reported in Figure 54.

RESULTS & DISCUSSION

Confusion Matrix

MaMic Dataset		
Normal/ Infected	Truth	
Predicted Value	1140	8
	3	1578

Hospital Dataset		
Normal/ Infected	Truth	
Predicted Value	741	8
	2	659

Combined Dataset		
Normal/ Infected	Truth	
Predicted Value	396	14
	4	906

Performance Statistics

Infection	Sensitivity	Specificity	Accuracy
MaMic	0.9973	0.9949	0.9959
Hospital	0.9973	0.9880	0.9929
Combination	0.9900	0.9847	0.9863
Average	**0.9948**	**0.9892**	**0.9917**

Figure 54. Confusion matrix and performance of AdaBoost Classifier across three dataset

8.3.2. Comparative Study

Analysis of thick and thin blood smear image for detection of Malaria parasite is a well tread domain over the last two decades. However, the complex staining of the red blood cells, Malaria parasite and white blood cell, non-uniform illumination makes it difficult to develop an automated algorithm for Malaria parasite detection that works across all dataset and all staining practices followed. In an attempt to extend the works of Diaz et al. [112], Linder et al. [165], Sio et al. [104], and Tek et al. [89], a method has been proposed in this paper that combines unsupervised learning techniques with rule and supervised learning based methods for optimal performance across a large array of datasets/databases. The method has been duly compared in terms of prediction efficiency (i.e. accuracy, sensitivity and specificity) and procedural differences with other state of the art methods proposed by other researchers. The comparative study has been elaborately documented in Table 21.

RESULTS & DISCUSSION

Authors	Year	Methods (Segmentation, Features, Classifications)	Dataset	Stage & Species Classification Parasitaemia	Overall Performance	Remarks
Nag et al (Proposed Method)	2018	Segmentation: Using Unsupervised clustering. Non-uniform background separation, Colour down-sampling and matching for suspect region identification, Candidate selection for de-clumping (Area based Threshold) Features: Morphological (**features introduced based on parasite and modified cell morphology**) and Textural features – 95/100 feature set Classification: k-NN, SVM, Naïve Bayes, Ensemble (combination of three weighted) and **Adaboost**	3 dataset (MaMic Dataset with 1410 images & Custom Dataset with 1320 images and a combined Dataset of 2730 images)	Stage Classification (Ring, Trophozoite, Schizont and Gametocyte – 4 stages) Species (P. vivax and P. falciparum) Prevalent in India. **RBC Enumeration and Parasitaemia Calculation**	Stage and Species Classification results Ensemble (SVM, 3-NN, Naïve Bayes) Method Accuracy: 0.9874 Sensitivity: 0.9925 Specificity: 0.9820 **Adaboost (57 trees) Accuracy: 0.9917 Sensitivity: 0.9948 Specificity: 0.9892**	A complete Hybrid system using Rule-based and ML methods for extraction of parasite region and Analysis.
Rajaraman et al [154]	2018	Segmentation: A level-set based algorithm to detect and segment the RBC; Classification: Convolution Neural Network has three convolutional layers and two fully connected layers	150 P. falciparum-infected and 50 healthy patients, Mobile Camera Images at 40X resolution. Total Images (13,779+ 13,779=27558)	Only Detection Performed	Cell / Patient Level Accuracy: 0.986, 0.959 Sensitivity: 0.981, 0.947 Specificity: 0.992, 0.972	Screening Detection Performed only at 40X
Gopakumar et al [155]	2018	Segmentation: Otsu Thresholding and Marker controlled Watershed Algorithm; Classification: Convolution Neural Network has two convolutional layers and two	Cultured P. falciparum blood samples were used to obtain 765 FoV for experiments	Only Detection Performed	Method B (Best Focussed Image) Sensitivity: 0.9891 Specificity: 0.9939	A stack of image taken at different focal length (FoV)

		fully connected layers				
Rosado et al [156]	2017	Segmentation: Adaptive Thresholding for segmentation of respective region Features: Geometric, colour and Texture Classification: Using SVM by altering the gamma and C values for each species and stage	566 images captured by Mobile Camera containing P.falciparum, P. Malariae and P.ovale	Tropozoites, Schizonts and Gametocytes of P.falciparum, P. Malariae and P.ovale	Sensitivity: 73.9-96.2 Specificity: 92.6-99.3	Separate Thresholding done to extract specific stage
Bibin et al [157]	2017	Segmentation: Level Set Method Features: Colour Histogram, coherence vector, GLCM and GLRLM and LBP for Texture Classification: Deep Belief Network input layer 484 nodes, 4 layers of hidden nodes (600 each) and 2 output nodes	Nine Slides and 630 non-overlapping RGB images	Only Detection Performed	Sensitivity: 97.6 Specificity: 95.92	Experiments performed on images provided by another author
Devi et al [158]	2017	Segmentation: Binarization using Otsu Threshold, morphological operations, Marker controlled Watershed for Clump Splitting Features:134 features including intensity based and Textural like GLCM, Gabor Classification: Hybrid classifiers; Combination of k-NN, SVM, Naïve Bayes and ANN in every possible combinations are compared	400 Thin Smear Slide Images with Leishman Staining. 1302 images obtained from slides	Differentiation between normal, P. falciparum (ring, Schizont), P. vivax (ring, ameboid ring, Schizont and Gametocyte) – 7 Class Problem	Accuracy 96.54 with Hybrid Classifier	Segmentation of WBC not taken into consideration, Enumeration, Parasitaemia calculation not performed
Park et al [159]	2016	Segmentation: Not performed Features: 23 Morphological features Classification: Fisher	Only RBC (centrifuged) and cultured with P. falciparum. Quantitative phased	Differentiation into Falciparum Ring, Trophozoite and Schizont	Sensitivity, Specificity, Accuracy of 3 Classifiers for 3 Classes are provided	Overall performance values not provided

RESULTS & DISCUSSION

		Discriminant Analysis, Logistic Regression and k-NN	spectroscopy images			
Widodo et al [160]	2015	Segmentation: Active contour Features: 1st Order Texture and 2nd Order GLCM Texture features Classification: SVM	120 images from CDC site	Differentiation of Stage performed (3-stage)	Individual Accuracy for each stage ranging between (85%-100%)	Specie differentiation not done, no clarity on dataset used.
Das et al [91]	2013	Segmentation: Foreground separation using Sobel and segmentation using Marker Controlled Watershed Features: 80 Texture features and 16 Morphological features (96 features reduced to 94 features) Classification: Naïve Bayes and SVM –	150 slides. Image number not specified in text	P. vivax and P. falciparum stage classifications performed. (6-Class) PV (ring, S,G) PF (Ring, G) and Normal	SVM Classifier (9 features subset) Sensitivity – 96.62% Specificity – 88.51%	The no of images and detailed segmentation process not present
Purwar et al [105]	2011	Segmentation: Using Chan-Vese method for edge detection followed by Morphological operations Features: Intensity Classification: Probabilistic k-Means Clustering	20 slides (10 infected and 10 normal). 500 images	Only Detection Performed	Comparative study between Automated and manual detection. (Overall Sensitivity 100%, Specificity 50-80%	Only intensity based feature is used. Overall performance is not present
Tek et al [89]	2010	Segmentation: Background separation with Area Granulometry and Intensity based Thresholding Features: 83 features computed based on morphology and Colour information Classification: k-NN compared with Fisher Discriminant Analysis and BPNN	Custom Dataset of 630 RGB images of 4 species of Plasmodium (vivax, falciparum, ovale and Malariae)	Detection, Stage and Specie differentiation with Parasitaemia calculation 20 Class – including non-parasite, 16 class (4 stageX4species) and 4 class	Overall Detection results Accuracy: 93.3 Sensitivity: 72.4 Specificity: 97.6 Parasitaemia results not provided	First complete system for Malaria parasite detection using ML

RESULTS & DISCUSSION

| Ruberto et al [93] | 2002 | Segmentation: Using Morphological operation and Area Granulometry. Watershed Transform for De-clumping Classification based on Colour similarity | Dataset information absent in literature | Detection and Stage differentiation performed only | Comparative study present. Among detection result with manual observations. | Uses Rule-based approach for detection |

Table 21. Comparative Study of the overall performance of the proposed Detection and Classification System with other comparable methods proposed by different Authors. (Only Journal Papers are considered for comparisons)

Table 22 documents the complexity analysis and execution time for the proposed method and other comparable methods.

	Devi et. al. [56]	Das et. al. [35]	Tek et. al. [15]	Das et. al. [16]	Rosado et. al. [54]	Nag et. al. (Proposed)
Segmentation	$O(n^2)$	$O(n^2)$	$O(n^2)$	$O(n^2)$	$O(n^2)$	$O(n^2)$
Feature Extraction	$O(n^2)$	$O(n^2)$	$O(n^2)$	$O(n^2)$	$O(n^2)$	$O(n^2)$
Feature Set Selection	$O(n^2)$	$O(n^2)$	-	$O(n^2)$	-	$O(n^2)$
Training + Cross Validation +Classification	$O(n\,d + k\,n) + O(n^2) + O(n\,d)$	$O(c*d)$	$O(n\,d + k\,n)$ [Cross Validation – not performed]	$O(n^2)+O(n\,d)$	$O(n^2)$ [Cross Validation- not performed]	$O(d\,n\,\log n)$
Execution Time						
Segmentation + Feature Extraction + Feature Set Selection	55.8265 secs	28.8891 secs	22.4532 secs	26.652 secs	39.8624 secs	29.2294 secs.
Training + Classification	60.2456 secs	15.2124 secs	37.2456 secs	23.657 secs	24.5529 secs	21.3696 secs.
Performance Analysis						
Sensitivity	0.9317	0.7374	0.7030	0.7942	0.7987	0.9948
Specificity	0.9460	0.8276	0.8286	0.6992	0.6601	0.9892
Accuracy	0.9416	0.7916	0.7811	0.7357	0.7112	0.9917

*Table 22. Comparative Study of the execution time of the proposed Detection and Classification System with other comparable methods proposed by different Authors. (Only Journal Papers are considered for comparisons). **n:** Number of Datapoints, **d:** Number of Features (Feature Dimension), **k:** Number of K value for nearest neighbour/cluster and **c:** Number of Classes*

RESULTS & DISCUSSION

As evidenced from Table 22, the proposed algorithm outperforms other state of the art algorithms developed for Parasitaemia estimation and stage, specie classification of malaria parasite in terms of Sensitivity and Specificity to elucidate the superiority of the developed algorithm over other state of the art methods in terms of Sensitivity and Specificity values, the concept of Multiobjective Optimization has been duly incorporated. Based on Figure 55, the non-dominated set of points were marked out by maximization of the Sensitivity & Specificity values for the algorithms compared.

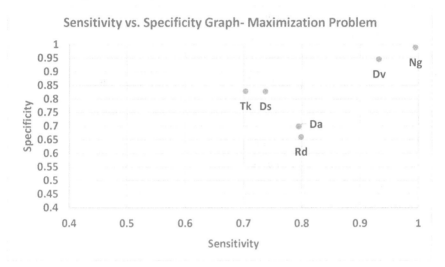

Figure 55: Representation of the algorithms that were compared on the same dataset based on the Sensitivity & Specificity values. Tk represents algorithm proposed by Tek et al [15], Ds represents the algorithm proposed by Das et al.[35], Da represents algorithm proposed by Das et al.[16], Rd represents algorithm proposed by Rosado et al.[54], Dv represents algorithm proposed by Devi et al.[56] and Ng represents algorithm proposed in this research work.

As a problem of Multiobjective Optimization, for a stable, reliable algorithm, the sensitivity as also the specificity value for its performance is expected to be close to 1. So the point with Sensitivity Specificity value of 1 is considered as the reference point in Figure X.

In view of Figure 55, the following dominance relations Ds ∥ Tk, Da ∥ Rd, Tk ∥ Da, Tk ∥ Rd, Ds ∥ Da, Ds ∥ Rd, Dv ≻ Tk, Dv ≻ Ds, Dv ≻ Da, Dv ≻ Rd, Ng ≻ Dv.were worked out. [where ≻ represents dominance and ∥ represents indifference/incomparability]

As Dominance Relation is transitive, hence Ng ≻ Tk, Ng ≻ Ds, Ng ≻ Da and Ng ≻ Rd

Thus based on Dominance Relation, the method prescribed/proposed dominates other state of the art methods in terms of Sensitivity & Specificity.

Additionally, the confusion matrix based on which the Sensitivity, Specificity and Accuracy values have been calculated are provided in Chapter 8 (Results And Discussion)

8.4. Summary

The contribution of the work is particularly two fold. In terms of application software it stands as a tool to assist medical practitioners at effective detection of Malaria parasite as also specie and stage classification. As opposed to other toolkits having similar functionality, this particular tool investigates parasites at cellular level which is much preferred by medical practitioners with significantly adding to the computational overhead. In perspective of computer science, it has been a long standing debate with regard to the predictive power of the classifiers. This paper adds on to the vast domain by putting forth a comparative study of single and ensemble classifiers (both inter and intra) using the same set of normalized filter features in perspective of the computer science domain.

As a result of 10 fold cross-validation, the final or best performance that has been achieved by the system is Accuracy of 0.9917, Sensitivity of 0.9948 and Specificity of 0.9892.

Like most other systems, the proposed model also suffers from certain limitations/drawbacks. Chapter 9 documents the limitations of the proposed model and suggests extension of the same for minimising such drawbacks.

CHAPTER 9
LIMITATION

9.1. Overview

The CAD systems that are designed and are practically used to assist medical practitioners in making informed decisions have their shortcomings and often stand erroneous in their prediction owing to large variability of biological systems, improper imaging equipment etc. The algorithm proposed has its limitations. Certain scenarios/image properties have been identified that greatly hinder the performance of the algorithm. These scenarios/properties are

- Presence of too many huge blood cell clumps in an image
- Low resolution/Poor Quality Image
- Image where cytoplasm is heavily stained.

9.2. Too many huge blood cell clumps

For an image where the number of blood cell clumps exceeds the number of singular red blood corpuscles, the de-clumping method was found to suffer, thereby making white blood cell segregation difficult. It affects Red Blood Cell enumeration and thereby affects Parasitaemia estimation. This happens because the median value is affected due to the overpowering presence of excessively large clusters. Figure 55 represents such a case where the number of clusters outnumbers single RBC cells present in the given image.

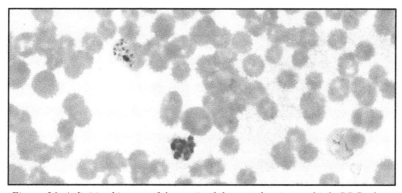

Figure 56. A digitized image of the aquired dataset showing multiple RBC clumps

A closer inspection of the digitized slide shall elucidate lysis of RBC due to pressure, environment etc. The distorted edges of the RBC further add to the complexity of the investigation process. The image has been obtained from the concerned partner hospital. Again, as per WHO requirements, the truncated RBC to the edges of the slide need to be removed for

LIMITATION

parasitaemia estimation. Ineffective de-clumping under such circumstances could actually lead to the removal of the entire clump including the infection at pre-processing stage, thereby hindering infection identification process.

To overcome this shortcoming de-clumping of all connected foreground components is required. In case of an image where the number of RBC outnumber the clumps, such a procedure is particularly time taking and tedious. However, when the number of clumps outnumber the number of single RBC, de-clumping of each connected component is cost effective as the number of connected components to be de-clumped are actually less. When the proposed algorithm was executed on an image taken from the slide represented in Figure 55, the de-clumping achieved have been duly represented in Figure 56.

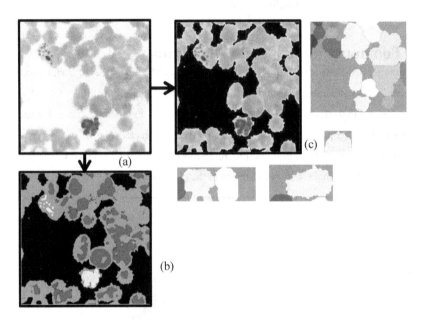

Figure 57. Declumping result of the image in Figure 54. (a) represents the original image, (b) represents the colored coded image and (C) represents the foreground segregated image

Figure 55a represents the original image, while Figure 55b represents the colored coded image and Figure 55c represents the foreground segregated image and the subsequent de-clumped sections. While in Figure 55b the problem areas are clearly marked out in green, the error in automated de-clumping hinders parasitaemia results. However, under most circumstances, owing to the fact that around 100 non-overlapping images are considered for each slide, hence such errors in de-clumping often lead to insignificant distortion in parasitaemia estimation. Another such example is demonstrated in Figure 57

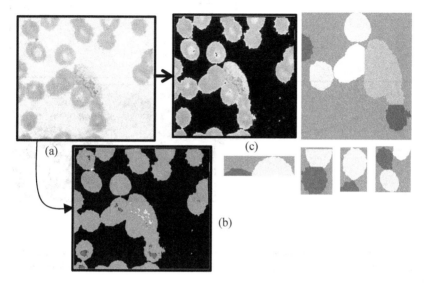

Figure 58. Another example of image where there are large number of RBC clumps affecting the declumping process

9.3. Low Resolution/Poor Quality Image

Apart from the presence of too many clumps, low resolution of images was also identified for limiting the performance of the algorithm in de-clumping. Figure 58 represents a heavily clumped blood smear image scanned at a resolution of 96 dpi. While the infection as also the artefacts can be neatly marked out and segregated before stage/specie classification, faulty declumping was found to hinder the trustworthiness of the estimated Parasitaemia value. The green marking in the extreme right end of Figure 58(c), marked in yellow, represents artifact while the green suspect region to the upper end of the image marked in yellow represent parasite. The green markings form part of the WBC Cluster. Texture features along with other morphological features (Table 7) were found to be effective at accurately differentiating infection from artifact and WBC nucleus.

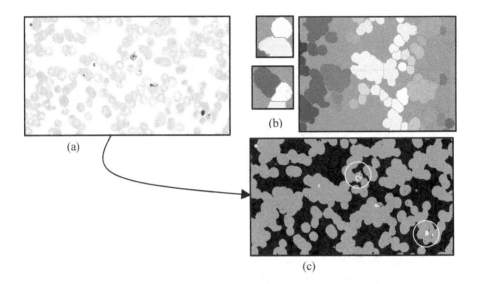

Figure 59. Showing a low resolution image with multiple clumps, (a) The image acquired from hospital. (b) The muliple coloured declumped image (c) Image marked with the infected regions.

9.4. Image where Cytoplasm is Heavily Stained

Lack in sharpness of image along with a heavily, unevenly stained cytoplasm was found to severely influence foreground particle identification by the proposed algorithm. Figure 59 documents such a case. In the thin blood smear image represented in Figure 59a, the cytoplasm is heavily and unevenly stained and takes in a blue hue. The edges of the RBCs are particularly undefined. The image has a resolution of only 72 dpi.

The image 58b, represents segregation of the foreground particles from the background, which is represented in black. Owing to the minimized colour difference in the foreground and the background, the unsupervised algorithm fails to segregate the foreground from the background even in the Lab colour space. Figure 59c represents the automated recoloring performed based on the YC_bC_r color model. The images represented in Figure 59b and 59c particularly highlight the shortcoming of the proposed algorithm in handling images that lack sharpness and in which the difference in the colour values of the foreground and the background is particularly low.

LIMITATION

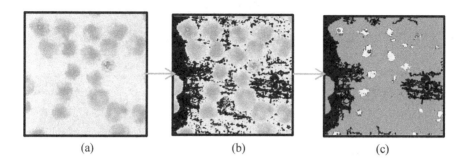

(a)   (b)   (c)

Figure 60. A highly stained image from the acquired datasetwhere, (a) is the original image containing a highly stained cytoplasm, (b) image showing the failure of background segregation and (c) image showing recoloured image that has overlapping background and foreground.

9.5. Summary

Based on the documentation, it can particularly be inferred that low resolution coupled with lack in sharpness and reduced difference in the foreground and background colour values is detrimental to the performance of the algorithm. However, most digitized images used for clinical examination under most circumstances do not suffer from this drawback. This makes the proposed algorithm applicable to real world scenario in clinical setting for aiding a medical practitioner in Malaria diagnosis.

A particular positive that is evident from the examples cited in this Chapter (i.e. Chapter 9) is that the algorithm is robust at handling differently stained thin blood smear images and can efficiently detect and classify Malaria parasite if present in the digitized smear image irrespective of the stain used.

Practical application of a proposed method stands only possible if it is cost effective and bears comparable costing, prediction performance to systems/methodology currently in use for achieving similar convenience. Chapter 10 provides a comparative study of the cost benefit analysis of the proposed algorithm against the methods in place for aiding the medical practitioner in Malaria parasite detection in thin blood smear image.

CHAPTER 10
COST ANALYSIS OF PROPOSED SYSTEM

10.1. Overview
At the very onset it is important to mention that the proposed CAD system was developed to assist the Pathologist/Medical Practitioner/Technicians and not to replace or substitute them. The idea of a CAD system for Histopathological Study originated with the availability of Whole Slide Scanners and Digital Microscope. Image digitization is the most important step in the process of CAD system for Malaria Parasite detection. Blood Smear is the simplest form of pathological tissue sample for histological analysis.

10.2. Present System
Automated Hematological Analyzers are present in most pathological laboratories. However, they can only quantify the components of blood and cannot detect the presence of parasites. Similarly, chemical assays and immunological tests can be used, however they are exorbitantly costly and quantification is not possible.

WHO considers Light Microscopy with Peripheral Blood Smear (PBS) as the 'Gold Standard'. This system is most effective and least costly method for parasite detection, stage and species identification. The only drawback of the system is that it is heavily dependent on manual labor (observation), subject to human bias and requires heuristic knowledge of the observer. The process involves:

- Step 1) Smear Slide Preparation: The process involving Step 1, includes collection of Blood sample (2 drops from finger prick) and preparation of smear slide, staining, drying and mounting of the slide.
- Step 2) Analysis of Slide under light microscope: This is done mostly by a trained microscopist or pathologist. Multiple views of non-overlapping slide positions are observed at 40X and 100X (Oil immersion lens) to predict the outcome of the tests.

10.3. Proposed System
The proposed System is dependent on conversion of the glass slide to a digital slide by using an image capturing hardware like whole slide scanners/digital microscope/or simply digital camera mounted on a conventional microscope. A set of non-overlapping digitized images are obtained from each slide. The images can thereafter analyzed, archived or transmitted over network (telemedicine). The images can be analyzed from anywhere, using PACS system or

can be fed to automated systems for batch processing. The output of such system/s is reports that will be subjected to verification by the pathologist. This reduces the workload of the concerned pathologist and more productivity for the laboratory.

- Step 1) Smear Slide Preparation: The process involving Step 1, includes collection of Blood sample (2 drops from finger prick) and preparation of smear slide, staining, drying and mounting of the slide.
- Step 2) Digital Slide Preparation: This is done mostly whole slide scanners at 40X and 100X resolution. Image set of each digital slide is put into folders.
- Step 3) Analysis of images by CAD system (multiple images of each slide in parallel processing mode)

10.4. Comparison of CAD System (Proposed) with Manual System (Existing)

In addition to infallibility, time taken for prediction; monetary cost, human labour cost etc. are factors that influence decision pertaining to whether a system can be used within a real world clinical setting or not. Table 23 documents the cost incurred against the benefits provided by the proposed system in contrast to manual blood smear slide investigation. Given that no digitized system is at present used at a clinical or commercial level to aid the medical practitioner in identification of Malaria parasite in blood smear image, hence the proposed algorithm cost has been compared to the manual system that is used as a standard by diagnosis laboratories for Malaria prediction.

COST ANALYSIS OF PROPOSED SYSTEM

Cost Heads	Present System		Proposed CAD	
	Resource in ₹	Time	Resource in ₹	Time
Slide Preparation	25/- (including Phlebologist and Slide Staining)	30 mins + Drying Time	25/- (including Phlebologist and Slide Staining)	30 mins + Drying Time
Digital Slide Preparation	-	-	One time Cost of Scanner (10L- 1 million slide Lifecycle) @ 10/- per slide (100+ image set)	15 mins/slide (including downtime) Batch processing of slides
Analysis of Image	**Technician Fee + Pathologist Fee** Tech: 30k/month & 30 slide/day @32/- per Slide Pathologist: 1.5L/month & 50 slides/day @100/- **Total 132/-/slide**	Technician (20 min/slide) + Pathologist (5 min/slide) + Processing time 10min **Total time 35 min/slide**	**CAD System + Pathologist** CAD System: (1L- 1 million image set Lifecycle) @ 1/- per slide + Overhead @1/- per slide = 2/- per Slide Pathologist: 1.5L/month & 100 slides/day @50/- (reduction due to 50% less workload) **Total 52/-/slide**	CAD System (3 min/slide due to Parallel processing in batch mode + Downtime) + Pathologist (3 min/slide on PACS system) + Processing time 1min **Total time 7 min/slide**
Total	Total 132/-/slide	Total time 65 min/slide	Total (77/-)/slide	Total time 48 min/slide
Benefit	Fully Manual so less of technology intervention		1. Automated Process with limited manual intervention by expert 2. Lower Cost 3. Ability to process in shorter time 4. Instant report generation 5. Image Archival 6. Possibility of Telemedicine facility (remote analysis by Pathologist) 7. No Human Bias and/or Fatigue	

Table 23. Table Showing Comparison of Cost Benefit Analysis of the Existing Manual Method and Automated CAD System (Proposed)

10.5. Summary

To implement or use CAD software as a trustworthy aid to a medical practitioner in disease diagnosis, the concerned system is required to undergo extensive testing for estimation of the prediction accuracy of the system, infallibility of the system which contributes toward the trustworthiness of the system. Additionally, the multiple tests conducted should be unbiased and globally accepted as a reliable benchmark/standard.

With this in view, Chapter 11 deals with an overview of the system developed, its positives and the extra step still required to enhance its infallibility and thereby make it a an aid medical practitioners can rely on and use in clinical investigation.

CHAPTER 11
CONCLUSION

11.1. Research Outcome

Though years of clinical study and research has been invested for eradication of Malaria in tropical countries, yet Malaria still remains a much dreaded disease particularly in the developing and underdeveloped parts of the world. Identification of the disease from digitized, thin blood smear images still stands as a challenge within computer scientists. The complexity of parasite identification lies in the fact that the parasite often mimics a white blood cell nucleus at an advanced stage or emulates a platelet at an early stage. Biological systems are particularly variable and on similar lines Malaria parasite formation often defies identical geometrical similarity to its kind.

Having mentioned the complexity of the problem, the proposed system is developed not only to detect the presence of Malaria infection in thin blood smear image but also to classify the same based on its stage and specie. The proposed algorithm can be divided into two broad segments, namely, initial screening phase and main detection and classification phase. The initial screening phase takes into consideration thin blood smear images at 40X magnification and estimates the presence/absence of Malaria parasite in the image under consideration. So the initial screening phase can be considered as a binary output system that predicts whether the blood smear image under contains infection or not. Based on the prediction of the initial screening phase the infection is further investigated in the main algorithm. However, clinical investigation normally considers 100X blood smear image as a reliable standard for Malaria parasite detection. Though the initial phase has been installed to enhance the efficiency of the proposed algorithm from the perspective of the computer scientist (in terms of time complexity), such work around was not particularly considered effective in real world Malaria investigation as True Negatives and False Positives are liable to have detrimental effect on infected and uninfected individuals alike.

However, to aid problem understanding, the consultant medical practitioner handpicked 250 images from the publicly available MaMic database based on which the initial screening algorithm was developed and accordingly tested. The initial screening phase is based on a hybrid algorithm and uses unsupervised K-Means (K=3) Clustering to segregate the White Blood Cell nucleus matter and suspected infection region from the RBC components. For initial stage infection, euler number based pixel matching was performed in suspected RBC

CONCLUSION

components and the suspect regions grouped with the WBC Cluster. Features such as number of red blood cells within a radius of 5 pixels calculated from the boundary point of the closed component in the WBC cluster that is farthest from the centre of the connected component, median length, median breadth, area, perimeter, eccentricity, presence/absence of nucleus material (Table 5) were used to identify or differentiate advanced stage Malaria parasite infection from White Blood cell component. On the 250 images the initial screening system recorded an accuracy, sensitivity and specificity value of 0.97826, 0.98 and 0.97656 respectively.

If infection was identified in the initial screening phase the image was further investigated in the main phase of the algorithm. In this phase, based on image pre-processing and subsequent segmentation, two clusters were worked out, namely, the infected RBC cluster and the WBC cluster. The infected RBC cluster consisted of only infection pertaining to the groups *Vivax* Ring, *Vivax* Tropozoite, *Falciparum* Tropozoite, *Falciparum* Ring, *Falciparum* Schizont, *Falciparum* Gametocyte (Male, Female), while the WBC cluster consisted of normal WBC cell, infection, artefacts and certain RBC cell outliers. Morphological as also texture features such as Gabor and Tamura features were used to mark out the specie and stage of the infection using single classifiers and an ensemble system. Based on the 50 feature set worked out by CMIM algorithm, Adaboost classifier with 57 trees was found to outperform all other concerned classifiers for either of the datasets (i.e. MaMic and Hospital Acquired Dataset) under consideration. The prediction Accuracy, Sensitivity, Specificity value obtained for the Mixed Dataset (consisting of 2730 images) was 0.9917, 0.9948, 0.9892 respectively.

The proposed algorithm was found comparable with other state of the art algorithms documented for Malaria parasite detection both in terms of performance and time complexity. The novelty of the proposed work lies in the fact that it is the first work in the domain of Malaria parasite detection and subsequent stage/specie classification that uses the combined efficiency of the CMIM algorithm along with the classification ability of the Adaboost classifier. An algorithm whose cost benefit has been evaluated against standard Laboratory practice also makes the proposed work unique and particularly distinct from other algorithms proposed towards stage/specie classification of Malaria parasite.

11.2. Future Work

Research should not remain confined as lines of text in the pages of journals or thesis work it should particularly serve for the wellbeing of humanity at large. To make the proposed model particularly viable for use in real world clinical setting, the model still needs further testing

CONCLUSION

based on globally accepted standards. Additionally there still remains scope for further improvement in the de-clumping algorithm by making it invariant to the presence of too many clumps, low resolution. As future work, it is believed that if the image segmentation phase pertaining to automatic segregation of the foreground connected components from the unevenly tinted background is improved, the efficiency of the algorithm shall surely be enhanced making the proposed algorithm particularly robust.

The methodology structure can particularly be extended to identify and investigate other diseases whose presence can be inferred from digitized thin blood smear images. The concerned researcher would need to analyze the relevant features for the concerned disease in due consultation with a pathologist or medical practitioner.

The proposed algorithm and the suggested future work is intended to minimize bias and increase the prediction accuracy of age old the Malaria parasite detection that is followed as a standard in diagnostic laboratory setting.

REFERENCES

[1]   M. García-Rojo, B. Blobel, and A. Laurinavicius, *Perspectives on Digital Pathology*. Amsterdam: IOS Press, 2012.

[2]   D. Treanor and B. Williams, "The Leeds Guide to Digital Pathology," *The Leeds Teaching Hospitals NHS, University of Leeds*, 2019. [Online]. Available: https://www.virtualpathology.leeds.ac.uk/Research/clinical/. [Accessed: 15-May-2019].

[3]   G. Bueno, M. M. Fernández-Carrobles, O. Deniz, and M. García-Rojo, "New Trends of Emerging Technologies in Digital Pathology," *Pathobiology*, vol. 83, no. 2–3, pp. 61–69, 2016.

[4]   L. Pantanowitz, A. Sharma, A. B. Carter, T. M. Kur, A. Sussman, and J. H. Saltz, "Twenty Years of Digital Pathology: An Overview of the Road Travelled, What is on the Horizon, and the Emergence of Vendor-Neutral Archives," in *Journal of pathology informatics*, 2018.

[5]   R. Singh, L. G. Chubb, L. Pantanowitz, and A. V Parwani, "Standardization in digital pathology: Supplement 145 of the DICOM standards," in *Journal of pathology informatics*, 2011.

[6]   T. W. Bauer and R. J. Slaw, "Validating Whole-Slide Imaging for Consultation Diagnoses in Surgical Pathology," *Arch. Pathol. Lab. Med.*, vol. 138, no. 11, pp. 1459–1465, May 2014.

[7]   M. Salto-Tellez, P. Maxwell, and P. Hamilton, "Artificial intelligence—the third revolution in pathology," *Histopathology*, vol. 74, no. 3, pp. 372–376, Feb. 2019.

[8]   B. S. Kakkilaya, "The Challenge of Malaria," 2018. [Online]. Available: https://www.Malariasite.com/challenge-of-Malaria/. [Accessed: 10-May-2019].

[9]   R. Neghina, I. Iacobiciu, A. M. Neghina, and I. Marincu, "Malaria, a Journey in Time: In Search of the Lost Myths and Forgotten Stories," *Am. J. Med. Sci.*, vol. 340, no. 6, pp. 492–498, Dec. 2010.

[10] T. D. Otto *et al.*, "Genomes of all known members of a Plasmodium subgenus reveal paths to virulent human Malaria," *Nat. Microbiol.*, vol. 3, no. 6, pp. 687–697, 2018.

[11] R. Carter and K. N. Mendis, "Evolutionary and Historical Aspects of the Burden of Malaria," *Clin. Microbiol. Rev.*, vol. 15, no. 4, pp. 564 LP – 594, Oct. 2002.

[12] R. Sallares, *Malaria and Rome: A History of Malaria in Ancient Italy*. Oxford: Oxford University Press, 2002.

[13] M. Ziegler, "Early use of the term 'Malaria,'" *Contagions*, 2014. [Online]. Available: https://contagions.wordpress.com/2014/08/07/early-use-of-the-term-Malaria/. [Accessed: 14-May-2019].

[14] S. Coll-Black, A. Bhushan, and K. Fritsch, "Integrating poverty and gender into health programs: A Sourcebook for health professionals," *Nurs. Health Sci.*, vol. 9, no. 4, pp. 246–253, Dec. 2007.

[15] S. W. Lindsay, M. Jawara, K. Paine, M. Pinder, G. E. L. Walraven, and P. M. Emerson, "Changes in house design reduce exposure to Malaria mosquitoes," *Trop. Med. Int. Heal.*, vol. 8, no. 6, pp. 512–517, Jun. 2003.

[16] R. Plant, *Indigenous peoples/ethnic minorities and poverty reduction : regional report*. Environment and Social Safeguard Division, Regional and Sustainable Development Dept.,

REFERENCES

Asian Development Bank, 2002.

[17] S. Russell, "The economic burden of illness for households in developing countries: a review of studies focusing on Malaria, tuberculosis, and human immunodeficiency virus/acquired immunodeficiency syndrome.," *Am. J. Trop. Med. Hyg.*, vol. 71, no. 2 Suppl, pp. 147–55, Aug. 2004.

[18] F. D. McCarthy, H. Wolf, and Y. Wu, "Malaria and growth." p. 1, 31-Mar-2000.

[19] J. Sachs and P. Malaney, "The economic and social burden of Malaria," *Nature*, vol. 415, no. 6872, pp. 680–685, Feb. 2002.

[20] World Health Organization, "WHO | This year's World Malaria report at a glance," *Who*, 2019. [Online]. Available: https://www.who.int/Malaria/media/world-Malaria-report-2018/en/#Global and regional Malaria burden, in numbers. [Accessed: 14-May-2019].

[21] WHO, "Malaria Key points : World Malaria report 2017," *World Health Organization*, 2018. [Online]. Available: https://www.who.int/Malaria/media/world-Malaria-report-2017/en/. [Accessed: 14-May-2019].

[22] WHO, "World Malaria Report 2016. Switzerland," *World Health Organization*, 2016. [Online]. Available: http://apps.who.int/iris/bitstream/10665/252038/1/9789241511711-eng.pdf?ua=1. [Accessed: 15-May-2019].

[23] D. of P. D. Global Health, "Parasites - Malaria," *CDC Govt. of USA*, 2019. [Online]. Available: https://www.cdc.gov/parasites/Malaria/index.html. [Accessed: 15-May-2019].

[24] D. of P. D. and M. Global Health, "DPDx - Laboratory Identification of Parasites of Public Health Concern - Malaria," *CDC Govt. of USA*, 2017. [Online]. Available: https://www.cdc.gov/dpdx/Malaria/index.html. [Accessed: 14-May-2019].

[25] K. Chatterjee, *Parasitology (protozoology and helminthology) in relation to clinical medicine*. CBC Publishers, 2009.

[26] S. Looareesuwan, C. J. Canfield, D. B. Hutchinson, P. Wilairatana, K. Chalermarut, and Y. Rattanapong, "Efficacy and safety of atovaquone/proguanil compared with mefloquine for treatment of acute Plasmodium falciparum Malaria in Thailand.," *Am. J. Trop. Med. Hyg.*, vol. 60, no. 4, pp. 526–532, Apr. 1999.

[27] D. J. Kyabayinze, J. K. Tibenderana, G. W. Odong, J. B. Rwakimari, and H. Counihan, "Operational accuracy and comparative persistent antigenicity of HRP2 rapid diagnostic tests for Plasmodium falciparum Malaria in a hyperendemic region of Uganda," *Malar. J.*, vol. 7, no. 1, p. 221, Dec. 2008.

[28] P. L. Bhandari, C. V Raghuveer, A. Rajeev, and P. D. Bhandari, "Comparative study of peripheral blood smear, quantitative buffy coat and modified centrifuged blood smear in Malaria diagnosis.," *Indian J. Pathol. Microbiol.*, vol. 51, no. 1, pp. 108–12, 2008.

[29] D. C. Warhurst and J. E. Williams, "ACP Broadsheet no 148. July 1996. Laboratory diagnosis of Malaria.," *J. Clin. Pathol.*, vol. 49, no. 7, pp. 533–8, Jul. 1996.

[30] A. R. Bharti *et al.*, "Polymerase chain reaction detection of Plasmodium vivax and Plasmodium falciparum DNA from stored serum samples: implications for retrospective diagnosis of Malaria.," *Am. J. Trop. Med. Hyg.*, vol. 77, no. 3, pp. 444–6, Sep. 2007.

[31] J. Storey, *Basic Malaria microscopy – Part I: Learner's guide*, 5th ed. Geneva: World Health Organization, 2010.

REFERENCES

[32] Y. S. Bahendwar and U. K. Chandra, "Detection of Malaria Parasites through Medical Image Segmentation Using ANN Algorithm," *Int. J. Adv. Res. Comput. Sci. Softw. Eng.*, vol. 5, no. 7, pp. 1063–1067, 2015.

[33] K. Chotivanich, K. Silamut, and N. P. J. Day, "Laboratory diagnosis of Malaria infection - A short review of methods," *New Zealand Journal of Medical Laboratory Science*, vol. 61, no. 1. pp. 4–7, 2007.

[34] D. Payne, "Use and limitations of light microscopy for diagnosing Malaria at the primary health care level," *Bull World Heal. Organ*, vol. 66, no. 5, pp. 621–626, 1988.

[35] C. Ohrt, Purnomo, M. A. Sutamihardja, D. Tang, and K. C. Kain, "Impact of Microscopy Error on Estimates of Protective Efficacy in Malaria-Prevention Trials," *J. Infect. Dis.*, vol. 186, no. 4, pp. 540–546, 2002.

[36] L. K. Erdman and K. C. Kain, "Molecular diagnostic and surveillance tools for global Malaria control," *Travel Med. Infect. Dis.*, vol. 6, no. 1–2, pp. 82–99, Jan. 2008.

[37] T. E. Clendennen, G. W. Long, and J. Kevin Baird, "QBC® and Giemsa-stained thick blood films: Diagnostic performance of laboratory technologists," *Trans. R. Soc. Trop. Med. Hyg.*, vol. 89, no. 2, pp. 183–184, Mar. 1995.

[38] A. Moody, "Rapid diagnostic tests for Malaria parasites," *Clin Microbiol Rev*, vol. 15, pp. 66–68, 2002.

[39] L. Ochola, P. Vounatsou, T. Smith, M. Mabaso, and C. Newton, "The reliability of diagnostic techniques in the diagnosis and management of Malaria in the absence of a gold standard," *Lancet Infect. Dis.*, vol. 6, no. 9, pp. 582–588, Sep. 2006.

[40] World Health Organization, "Malaria Rapid Diagnostic Test Performance, Summary results of WHO product testing of Malaria RDTs: rounds 1-6 (2008–2015)," Geneva, 2015.

[41] S. W. Lee, K. Jeon, B. R. Jeon, and I. Park, "Rapid diagnosis of vivax Malaria by the SD Bioline Malaria Antigen test when thrombocytopenia is present.," *J. Clin. Microbiol.*, vol. 46, no. 3, pp. 939–42, Mar. 2008.

[42] T. S. Park et al., "Diagnostic Usefulness of SD Malaria Antigen and Antibody Kits for Differential Diagnosis of vivax Malaria in Patients with Fever of Unknown Origin," *Korean J. Lab. Med.*, vol. 26, no. 4, p. 241, Aug. 2009.

[43] S. H. Kim et al., "Evaluation of a rapid diagnostic test specific for Plasmodium vivax," *Trop. Med. Int. Heal.*, vol. 13, no. 12, pp. 1495–1500, Dec. 2008.

[44] T. F. McCutchan, R. C. Piper, and M. T. Makler, "Use of Malaria rapid diagnostic test to identify Plasmodium knowlesi infection," *Emerg. Infect. Dis.*, vol. 14, no. 11, pp. 1750–1752, Nov. 2008.

[45] W. H. Organization and others, "World Malaria report 2014. 2014," World Health Organization, Geneva, 2015.

[46] M. M.L., M. M.I., A. S.M.K., K. E., and K. S.P., "Challenges in routine implementation and quality control of rapid diagnostic tests for Malaria-Rufiji District, Tanzania," *Am. J. Trop. Med. Hyg.*, vol. 79, no. 3, pp. 385–390, Sep. 2008.

[47] A. Ratsimbasoa, L. Fanazava, R. Radrianjafy, J. Ramilijaona, H. Rafanomezantsoa, and D. Ménard, "Short report: Evaluation of two new immunochromatographic assays for diagnosis of Malaria," *Am. J. Trop. Med. Hyg.*, vol. 79, no. 5, pp. 670–672, Nov. 2008.

REFERENCES

[48] T. H., B. J., B. E., G. B., and C. D., "Performance of the OptiMAL® dipstick in the diagnosis of Malaria infection in pregnancy," *Ther. Clin. Risk Manag.*, vol. 4, no. 3, pp. 631–636, Jun. 2008.

[49] L. K. Erdman and K. C. Kain, "Molecular diagnostic and surveillance tools for global Malaria control," *Travel Med. Infect. Dis.*, vol. 6, no. 1–2, pp. 82–99, Jan. 2008.

[50] R. C. She, M. L. Rawlins, R. Mohl, S. L. Perkins, H. R. Hill, and C. M. Litwin, "Comparison of immunofluorescence antibody testing and two enzyme immunoassays in the serologic diagnosis of Malaria," *J. Travel Med.*, vol. 14, no. 2, pp. 105–111, Mar. 2007.

[51] C. Doderer et al., "A new ELISA kit which uses a combination of Plasmodium falciparum extract and recombinant Plasmodium vivax antigens as an alternative to IFAT for detection of Malaria antibodies," *Malar. J.*, vol. 6, p. 19, Feb. 2007.

[52] H. W. Reesink, "European strategies against the parasite transfusion risk," *Transfus. Clin. Biol.*, vol. 12, no. 1, pp. 1–4, Feb. 2005.

[53] M. Mungai, G. Tegtmeier, M. Chamberland, and M. Parise, "Transfusion-Transmitted Malaria in the United States from 1963 through 1999," *N. Engl. J. Med.*, vol. 344, no. 26, pp. 1973–1978, Jun. 2002.

[54] A. J. Sulzer, M. Wilson, and E. C. Hall, "Indirect fluorescent-antibody tests for parasitic diseases. V. An evaluation of a thick-smear antigen in the IFA test for Malaria antibodies.," *Am. J. Trop. Med. Hyg.*, vol. 18, no. 2, pp. 199–205, Mar. 1969.

[55] B. Morassin, R. Fabre, A. Berry, and J. F. Magnaval, "One year's experience with the polymerase chain reaction as a routine method for the diagnosis of imported Malaria," *Am. J. Trop. Med. Hyg.*, vol. 66, no. 5, pp. 503–508, May 2002.

[56] M. T. Makler, C. J. Palmer, and A. L. Ager, "A review of practical techniques for the diagnosis of Malaria," *Ann. Trop. Med. Parasitol.*, vol. 92, no. 4, pp. 419–433, Jun. 1998.

[57] H. Swan et al., "Evaluation of a real-time polymerase chain reaction assay for the diagnosis of Malaria in patients from Thailand.," *Am. J. Trop. Med. Hyg.*, vol. 73, no. 5, pp. 850–4, Nov. 2005.

[58] T. Hänscheid and M. P. Grobusch, "How useful is PCR in the diagnosis of Malaria?," *Trends Parasitol.*, vol. 18, no. 9, pp. 395–8, Sep. 2002.

[59] P. F. Mens, A. van Amerongen, P. Sawa, P. A. Kager, and H. D. F. H. Schallig, "Molecular diagnosis of Malaria in the field: development of a novel 1-step nucleic acid lateral flow immunoassay for the detection of all 4 human Plasmodium spp. and its evaluation in Mbita, Kenya," *Diagn. Microbiol. Infect. Dis.*, vol. 61, no. 4, pp. 421–427, Aug. 2008.

[60] L. L. M. Poon et al., "Sensitive and inexpensive molecular test for falciparum Malaria: Defecting Plasmodium falciparum DNA directly from heat-treated blood by loop-mediated isothermal amplification," *Clin. Chem.*, vol. 52, no. 2, pp. 303–306, Nov. 2006.

[61] E. T. Han et al., "Detection of four Plasmodium species by genus- and species-specific loop-mediated isothermal amplification for clinical diagnosis," *J. Clin. Microbiol.*, vol. 45, no. 8, pp. 2521–2528, Aug. 2007.

[62] H. Aonuma et al., "Rapid identification of Plasmodium-carrying mosquitoes using loop-mediated isothermal amplification," *Biochem. Biophys. Res. Commun.*, vol. 376, no. 4, pp. 671–676, Nov. 2008.

[63] V. Wongchotigul et al., "The use of flow cytometry as a diagnostic test for Malaria parasites," *Southeast Asian J. Trop. Med. Public Health*, vol. 35, no. 3, pp. 552–559, Sep. 2004.

REFERENCES

[64] M. P. Grobusch *et al.*, "Sensitivity of Hemozoin Detection by Automated Flow Cytometry in Non- and Semi-Immune Malaria Patients," *Cytom. Part B - Clin. Cytom.*, vol. 55, no. 1, pp. 46–51, Sep. 2003.

[65] A. J. De Langen, J. Van Dillen, P. De Witte, S. Mucheto, N. Nagelkerke, and P. Kager, "Automated detection of Malaria pigment: Feasibility for Malaria diagnosing in an area with seasonal Malaria in northern Namibia," *Trop. Med. Int. Heal.*, vol. 11, no. 6, pp. 809–816, Jun. 2006.

[66] M. Kotepui, K. Uthaisar, B. Phunphuech, and N. Phiwklam, "A diagnostic tool for Malaria based on computer software," *Sci. Rep.*, vol. 5, no. 1, p. 16656, Dec. 2015.

[67] C. Briggs *et al.*, "Development of an Automated Malaria Discriminant Factor Using VCS Technology," *Am. J. Clin. Pathol.*, vol. 126, no. 5, pp. 691–698, Nov. 2006.

[68] B. V. Mendelow *et al.*, "Automated Malaria detection by depolarization of laser light," *Br. J. Haematol.*, vol. 104, no. 3, pp. 499–503, Mar. 1999.

[69] V. Mikhailovich, D. Gryadunov, A. Kolchinsky, A. A. Makarov, and A. Zasedatelev, "DNA microarrays in the clinic: Infectious diseases," *BioEssays*, vol. 30, no. 7, pp. 673–682, Jul. 2008.

[70] G. Palacios *et al.*, "Panmicrobial Oligonucleotide Array for Diagnosis of Infectious Diseases," *Emerg. Infect. Dis.*, vol. 13, no. 1, pp. 73–81, Jan. 2007.

[71] S. P.F. *et al.*, "Rapid detection of Malaria infection in vivo by laser desorption mass spectrometry," *Am. J. Trop. Med. Hyg.*, vol. 71, no. 5, pp. 546–551, Nov. 2004.

[72] N. Tangpukdee, C. Duangdee, P. Wilairatana, and S. Krudsood, "Malaria diagnosis: A brief review," *Korean J. Parasitol.*, vol. 47, no. 2, pp. 93–102, 2009.

[73] T. Hänscheid, "Diagnosis of Malaria: a review of alternatives to conventional microscopy.," *Clin. Lab. Haematol.*, vol. 21, no. 4, pp. 235–45, Aug. 1999.

[74] C. Wongsrichanalai, M. J. Barcus, S. Muth, A. Sutamihardja, and W. H. Wernsdorfer, "A review of Malaria diagnostic tools: microscopy and rapid diagnostic test (RDT).," *Am. J. Trop. Med. Hyg.*, vol. 77, no. 6 Suppl, pp. 119–27, Dec. 2007.

[75] F. B. Tek, A. G. Dempster, and I. Kale, "Computer vision for microscopy diagnosis of Malaria," *Malar. J.*, vol. 8, no. 1, p. 153, Dec. 2009.

[76] J. Frean, *Microscopic determination of Malaria parasite load: Role of image analysis*, vol. 3. 2010.

[77] E. A. Mohammed, M. M. Mohamed, B. H. Far, and C. Naugler, "Peripheral blood smear image analysis: A comprehensive review," *J Pathol Inf.*, vol. 5, no. 1, p. 9, 2014.

[78] U. K. Chandra and Y. Bahendwar, "Review on CAD based System for Detection of Disease through Medical Image Processing," *Rev. CAD based Syst. Detect. Dis. through Med. Image Process.*, vol. 4, pp. 285–290, 2015.

[79] Z. Jan, A. Khan, M. Sajjad, K. Muhammad, S. Rho, and I. Mehmood, "A review on automated diagnosis of Malaria parasite in microscopic blood smears images," *Multimed. Tools Appl.*, vol. 77, no. 8, pp. 9801–9826, Apr. 2018.

[80] L. Rosado, J. M. Correia da Costa, D. Elias, and J. Cardoso, *A Review of Automatic Malaria Parasites Detection and Segmentation in Microscopic Images*, vol. 14. 2016.

REFERENCES

[81] A. Loddo, C. Di Ruberto, and M. Kocher, "Recent Advances of Malaria Parasites Detection Systems Based on Mathematical Morphology," *Sensors*, vol. 18, no. 2, p. 513, Feb. 2018.

[82] M. Poostchi, K. Silamut, R. J. Maude, S. Jaeger, and G. Thoma, "Image analysis and machine learning for detecting Malaria," *Transl. Res.*, vol. 194, pp. 36–55, Apr. 2018.

[83] S. S. Devi, S. Alam Sheikh, and R. Laskar, *Erythrocyte Features for Malaria Parasite Detection in Microscopic Images of Thin Blood Smear: A Review*, vol. 4. 2016.

[84] D. K. DAS, R. MUKHERJEE, and C. CHAKRABORTY, "Computational microscopic imaging for Malaria parasite detection: a systematic review," *J. Microsc.*, vol. 260, no. 1, pp. 1–19, Oct. 2015.

[85] M. J. Sadiq and V. V. S. S. S. Balaram, "Review of Microscopic Image Processing Techniques to wards Malaria Infected Erythrocyte Detection from Thin Blood Smears," *Glob. J. Comput. Sci. Technol. F Graph. Vis.*, vol. 17, no. 2, 2017.

[86] M. J. Sadiq and V. V. S. S. S. Balaram, "Review of Microscopic Image Processing techniques towards Malaria Infected Erythrocyte Detection from Thin Blood Smears," *Int. J. Sci. Eng. Res.*, vol. 8, no. 6, pp. 1633–1638, 2017.

[87] C. Jagtap D and U. Rani N, "The Review o f Microscopic Image Analysis a nd Computer Aided Malaria Parasite Infected Erythrocyte Detection," *Int. J. Comput. Intell. Res.*, vol. 13, no. 4, pp. 485–495, 2017.

[88] W. A. Saputra, H. A. Nugroho, and A. E. Permanasari, "Toward development of automated plasmodium detection for Malaria diagnosis in thin blood smear image: An overview," in *2016 International Conference on Information Technology Systems and Innovation (ICITSI)*, 2016, pp. 1–6.

[89] F. B. Tek, A. G. Dempster, and İ. Kale, "Parasite detection and identification for automated thin blood film Malaria diagnosis," *Comput. Vis. Image Underst.*, vol. 114, no. 1, pp. 21–32, 2010.

[90] F. B. Tek, A. G. Dempster, and I. Kale, "Malaria Parasite Detection in Peripheral Blood Images," in *Proceedings of the British Machine Vision Conference 2006*, 2006, pp. 36.1-36.10.

[91] D. K. Das, M. Ghosh, M. Pal, A. K. Maiti, and C. Chakraborty, "Machine learning approach for automated screening of Malaria parasite using light microscopic images," *Micron*, vol. 45, pp. 97–106, 2013.

[92] E. Y. Lam, "Combining gray world and retinex theory for automatic white balance in digital photography," in *Proceedings of the Ninth International Symposium on Consumer Electronics, 2005. (ISCE 2005).*, 2005, pp. 134–139.

[93] C. Di Ruberto, A. Dempster, S. Khan, and B. Jarra, "Analysis of infected blood cell images using morphological operators," *Image Vis. Comput.*, vol. 20, no. 2, pp. 133–146, 2002.

[94] N. E. Ross, C. J. Pritchard, D. M. Rubin, and A. G. Dusé, "Automated image processing method for the diagnosis and classification of Malaria on thin blood smears," *Med. Biol. Eng. Comput.*, vol. 44, no. 5, pp. 427–436, 2006.

[95] D. Anggraini, A. S. Nugroho, C. Pratama, I. E. Rozi, A. A. Iskandar, and R. N. Hartono, "Automated status identification of microscopic images obtained from Malaria thin blood smears," in *Proceedings of the 2011 International Conference on Electrical Engineering and Informatics*, 2011, pp. 1–6.

[96] L. Rosado, J. M. C. da Costa, D. Elias, and J. S. Cardoso, "Automated Detection of Malaria

REFERENCES

Parasites on Thick Blood Smears via Mobile Devices," *Procedia Comput. Sci.*, vol. 90, pp. 138–144, 2016.

[97] W. Preedanan, M. Phothisonothai, W. Senavongse, and S. Tantisatirapong, "Automated detection of plasmodium falciparum from Giemsa-stained thin blood films," in *2016 8th International Conference on Knowledge and Smart Technology (KST)*, 2016, pp. 215–218.

[98] Y. S. Bahendwar and U. K. Chandra, "Detection of Malaria Parasites through Medical Image Segmentation Using ANN Algorithm," *Int. J. Adv. Res. Comput. Sci. Softw. Eng.*, vol. 5, pp. 1063–1067, 2015.

[99] V. V. Makkapati and R. M. Rao, "Segmentation of Malaria parasites in peripheral blood smear images," in *ICASSP, IEEE International Conference on Acoustics, Speech and Signal Processing - Proceedings*, 2009, pp. 1361–1364.

[100] L. Gitonga, D. Maitethia Memeu, K. Kaduki, M. Allen Christopher Kale, and N. Muriuki, *Determination of Plasmodium Parasite Life Stages and Species in Images of Thin Blood Smears Using Artificial Neural Network*, vol. 04. 2014.

[101] H. A. Nugroho, S. A. Akbar, and E. E. H. Murhandarwati, "Feature extraction and classification for detection Malaria parasites in thin blood smear," in *2015 2nd International Conference on Information Technology, Computer, and Electrical Engineering (ICITACEE)*, 2015, pp. 197–201.

[102] I. R. Dave and K. P. Upla, "Computer aided diagnosis of Malaria disease for thin and thick blood smear microscopic images," in *2017 4th International Conference on Signal Processing and Integrated Networks (SPIN)*, 2017, pp. 561–565.

[103] S. S. Savkare and S. P. Narote, "Automated system for Malaria parasite identification," in *2015 International Conference on Communication, Information & Computing Technology (ICCICT)*, 2015, pp. 1–4.

[104] S. W. S. Sio *et al.*, "MalariaCount: An image analysis-based program for the accurate determination of parasitemia," *J. Microbiol. Methods*, vol. 68, no. 1, pp. 11–18, 2007.

[105] Y. Purwar, S. L. Shah, G. Clarke, A. Almugairi, and A. Muehlenbachs, "Automated and unsupervised detection of Malarial parasites in microscopic images," *Malar. J.*, vol. 10, no. 1, p. 364, 2011.

[106] J. Somasekar and B. Eswara Reddy, "Segmentation of erythrocytes infected with Malaria parasites for the diagnosis using microscopy imaging," *Comput. Electr. Eng.*, vol. 45, pp. 336–351, Jul. 2015.

[107] J. E. Arco, J. M. Górriz, J. Ramírez, I. Álvarez, and C. G. Puntonet, "Digital image analysis for automatic enumeration of Malaria parasites using morphological operations," *Expert Syst. Appl.*, vol. 42, no. 6, pp. 3041–3047, 2015.

[108] J.Somasekar, B. E. Reddy, E. K. Reddy, and C.-H. Lai, "An Image Processing Approach for Accurate Determination of Parasitemia in Peripheral Blood Smear Images," *IJCA Spec. Issue Nov. Asp. Digit. Imaging Appl.*, no. 1, pp. 23–28, 2011.

[109] N. Ahirwar, S. Pattnaik, and B. Acharya, *Advanced image analysis based system for automatic detection and classification Malarial parasite in blood images*, vol. 5. 2012.

[110] M. I. Khan, B. K. Singh, B. Acharya, and J. Soni, "Content Based Image Retrieval Approaches for Detection of Malarial in Blood Images," *Int. J. Biometrics Bioinforma.*, vol. 5, no. 2, pp. 97–110, 2011.

[111] S. K. Reni, I. Kale, and R. Morling, "Analysis of thin blood images for automated Malaria

REFERENCES

diagnosis," in *2015 E-Health and Bioengineering Conference (EHB)*, 2015, pp. 1–4.

[112] G. Díaz, F. A. González, and E. Romero, "A semi-automatic method for quantification and classification of erythrocytes infected with Malaria parasites in microscopic images," *J. Biomed. Inform.*, vol. 42, no. 2, pp. 296–307, 2009.

[113] C. Di Rubeto, A. Dempster, S. Khan, and B. Jarra, "Segmentation of blood images using morphological operators," in *Proceedings 15th International Conference on Pattern Recognition. ICPR-2000*, 2002, vol. 3, pp. 397–400.

[114] M. I. Khan, B. K. Singh, B. Acharya, and J. Soni, "Content Based Image Retrieval Approaches for Detection of Malarial Parasite in Blood Images ," *Clin. Microbiol. Rev.*, vol. 15, no. 2, pp. 97–110, 2011.

[115] S. Kareem, I. Kale, and R. C. S. Morling, "Automated Malaria parasite detection in thin blood films:-A hybrid illumination and color constancy insensitive, morphological approach," in *IEEE Asia-Pacific Conference on Circuits and Systems, Proceedings, APCCAS*, 2012, pp. 240–243.

[116] S. Kareem, R. C. S. Morling, and I. Kale, "A novel method to count the red blood cells in thin blood films," in *Proceedings - IEEE International Symposium on Circuits and Systems*, 2011, pp. 1021–1024.

[117] S. Kumar, S. H. Ong, S. Ranganath, T. C. Ong, and F. T. Chew, "A Rule-based Approach for Robust Clump Splitting," *Pattern Recogn.*, vol. 39, no. 6, pp. 1088–1098, Jun. 2006.

[118] V. K. Bairagi and K. C. Charpe, "Comparison of Texture Features Used for Classification of Life Stages of Malaria Parasite.," *Int. J. Biomed. Imaging*, vol. 2016, p. 7214156, 2016.

[119] K. Prasad, J. Winter, U. M. Bhat, R. V Acharya, and G. K. Prabhu, "Image analysis approach for development of a decision support system for detection of Malaria parasites in thin blood smear images," *J. Digit. Imaging*, vol. 25, no. 4, pp. 542–549, Aug. 2012.

[120] A. Mehrjou, T. Abbasian, and M. Izadi, "Automatic Malaria Diagnosis system," in *2013 First RSI/ISM International Conference on Robotics and Mechatronics (ICRoM)*, 2013, pp. 205–211.

[121] D. Das, M. Ghosh, C. Chakraborty, A. K. Maiti, and M. Pal, "Probabilistic prediction of Malaria using morphological and textural information," in *2011 International Conference on Image Information Processing*, 2011, pp. 1–6.

[122] K. Zuiderveld, "Contrast Limited Adaptive Histogram Equalization," in *Graphics Gems*, AP Professional, 2013, pp. 474–485.

[123] J. A. Frean, "Reliable enumeration of Malaria parasites in thick blood films using digital image analysis," *Malar. J.*, vol. 8, no. 1, p. 218, Dec. 2009.

[124] P. Ghosh, D. Bhattacharjee, M. Nasipuri, and D. K. Basu, *Medical aid for automatic detection of Malaria*, vol. 245 CCIS. 2011.

[125] J. Somsekar, "Computer Vision for Malaria Parasite Classification in Erythrocytes," *Int. J. Comput. Sci. Eng.*, vol. 3, no. 6, pp. 2251–2256, 2011.

[126] L. Damahe, N. Thakur, R. K. Krishna, and N. Janwe, *Segmentation Based Approach to Detect Parasites and RBCs in Blood Cell Images*, vol. 4. 2011.

[127] G. W. Zack, W. E. Rogers, and S. A. Latt, "Automatic measurement of sister chromatid exchange frequency.," *J. Histochem. Cytochem.*, vol. 25, no. 7, pp. 741–753, Jul. 1977.

[128] L. Wei, L. Y. Sheng, R. X. Yi, and D. Peng, "A new approach for extracting the contour of an ROI

in medical images," in *Proceedings - 2008 International Conference on Advanced Computer Theory and Engineering, ICACTE 2008*, 2008, pp. 1025–1029.

[129] C. Di Ruberto, A. Dempster, S. Khan, and B. Jarra, "Automatic thresholding of infected blood images using granulometry and regional extrema," in *Proceedings 15th International Conference on Pattern Recognition. ICPR-2000*, 2002, vol. 3, pp. 441–444.

[130] A. Kumar, P. A. Choudhary, P. P. U. Tembhare, and P. C. R. Pote, "Enhanced Identification of Malarial Infected Objects using Otsu Algorithm from Thin Smear Digital Images," *Int. J. Latest Res. Sci. Technol.*, vol. 1, no. 2, pp. 159–163, 2012.

[131] E. Komagal, K. Sasi kumar, and A. Vigneswaran, "Recognition And Classification Of Malaria Plasmodium Diagnosis," *Int. J. Eng. Res. Technol.*, vol. 2, no. 1, pp. 1–5, 2013.

[132] P. T. Suradka, "Detection of Malarial Parasite in Blood Using Image Processing," *Int. J. Eng. Innov. Technol.*, vol. 2, pp. 124–126, 2013.

[133] V. Parkhi, P. Pawar, and A. Surve, "Computer Automation for Malaria Parasite Detection Using Linear Programming," *Int. J. Adv. Res. Electr. Electron. Instrum. Eng.*, vol. 2, no. 5, pp. 1984–1988, 2013.

[134] A. M. D. Ghate, "Automatic Detection of Malaria Parasite from Blood Images," *Int. J. Adv. Comput. Technol.*, vol. 4, no. 1, pp. 129–132, 2014.

[135] K. Chakraborty, "A Combined Algorithm for Malaria Detection from Thick Smear Blood Slides," *J. Heal. Med. Informatics*, vol. 06, no. 01, pp. 1–6, 2015.

[136] Yi-Wei Yu and Jung-Hua Wang, "Image segmentation based on region growing and edge detection," in *IEEE SMC'99 Conference Proceedings. 1999 IEEE International Conference on Systems, Man, and Cybernetics (Cat. No.99CH37028)*, vol. 6, pp. 798–803.

[137] S. Halim, T. R. Bretschneider, Y. Li, P. R. Preiser, and C. Kuss, "Estimating Malaria Parasitaemia from Blood Smear Images," in *2006 9th International Conference on Control, Automation, Robotics and Vision*, 2006, pp. 1–6.

[138] S. F. Toha and U. K. Ngah, "Computer Aided Medical Diagnosis for the Identification of Malaria Parasites," in *2007 International Conference on Signal Processing, Communications and Networking*, 2007, pp. 521–522.

[139] Y. Yuming Fang, W. Wei Xiong, W. Weisi Lin, and Z. Zhenzhong Chen, "Unsupervised Malaria parasite detection based on phase spectrum," in *2011 Annual International Conference of the IEEE Engineering in Medicine and Biology Society*, 2011, vol. 2011, pp. 7997–8000.

[140] M. Elter, E. Hasslmeyer, and T. Zerfass, "Detection of Malaria parasites in thick blood films," in *2011 Annual International Conference of the IEEE Engineering in Medicine and Biology Society*, 2011, vol. 2011, pp. 5140–5144.

[141] S. Raviraja, G. Bajpai, and S. K. Sharma, "Analysis of Detecting the Malarial Parasite Infected Blood Images Using Statistical Based Approach," Springer, Berlin, Heidelberg, 2007, pp. 502–505.

[142] N. A. Khan, H. Pervaz, A. K. Latif, A. Musharraf, and Saniya, "Unsupervised identification of Malaria parasites using computer vision," in *2014 11th International Joint Conference on Computer Science and Software Engineering (JCSSE)*, 2014, pp. 263–267.

[143] S. Annaldas and S. S. Shirgan, "Automatic Diagnosis of Malaria Parasites Using Neural Network and Support Vector Machine," *Int. J. Adv. Found. Res. Comput.*, vol. 2, pp. 60–66, 2015.

REFERENCES

[144] K. M. Rao, *Application of Mathematical Morphology to Biomedical Image Processing*. 2004.

[145] S. S. Savkare and S. P. Narote, *Automatic Detection of Malaria Parasites for Estimating Parasitemia*, vol. 2, no. 12. 2011.

[146] N. Otsu, "A Threshold Selection Method from Gray-Level Histograms," *IEEE Trans. Syst. Man. Cybern.*, vol. 9, no. 1, pp. 62–66, Jan. 1979.

[147] S. M. Smith and J. M. Brady, "SUSAN - A new approach to low level image processing," *Int. J. Comput. Vis.*, vol. 23, no. 1, pp. 45–78, 1997.

[148] A. S. Abdul Nasir, M. Y. Mashor, and Z. Mohamed, "Segmentation based approach for detection of Malaria parasites using moving k-means clustering," in *2012 IEEE-EMBS Conference on Biomedical Engineering and Sciences*, 2012, pp. 653–658.

[149] A. S. Abdul-Nasir, M. Y. Mashor, and Z. Mohamed, *Colour image segmentation approach for detection of Malaria parasites using various colour models and k-means clustering*, vol. 10, no. 1. 2013.

[150] S. Suryawanshi and V. V Dixit, "Comparative Study of Malaria Parasite Detection using Euclidean Distance Classifier & SVM," *Int. J. Adv. Res. Comput. Eng. Technol.*, vol. 2, no. 11, pp. 2994–2997, 2013.

[151] D. A. Kurer and V. P. Gejji, "Detection of Malarial Parasites in Blood Images," *Int. J. Eng. Sci. Innov. Technol.*, vol. 3, no. 3, pp. 651–656, 2014.

[152] M. L. Chayadevi and G. T. Raju, "Usage of ART for Automatic Malaria Parasite Identification Based on Fractal Features," *Int. J. Video &Image Process. Netw. Secur. IJVIPNS-IJENS*, vol. 14, no. 04, pp. 7–15, 2014.

[153] M. I. Razzak and B. Alhaqbani, "Automatic Detection of Malarial Parasite Using Microscopic Blood Images," *J. Med. Imaging Heal. Informatics*, vol. 5, no. 3, pp. 591–598, 2015.

[154] S. Rajaraman *et al.*, "Pre-trained convolutional neural networks as feature extractors toward improved Malaria parasite detection in thin blood smear images," *PeerJ*, vol. 6, p. e4568, Apr. 2018.

[155] G. P. Gopakumar, M. Swetha, G. Sai Siva, and G. R. K. Sai Subrahmanyam, "Convolutional neural network-based Malaria diagnosis from focus stack of blood smear images acquired using custom-built slide scanner," *J. Biophotonics*, vol. 11, no. 3, p. e201700003, Mar. 2018.

[156] L. Rosado, J. da Costa, D. Elias, and J. Cardoso, "Mobile-Based Analysis of Malaria-Infected Thin Blood Smears: Automated Species and Life Cycle Stage Determination," *Sensors*, vol. 17, no. 10, p. 2167, Sep. 2017.

[157] D. Bibin, M. S. Nair, and P. Punitha, "Malaria Parasite Detection From Peripheral Blood Smear Images Using Deep Belief Networks," *IEEE Access*, vol. 5, pp. 9099–9108, 2017.

[158] S. S. Devi, R. H. Laskar, and S. A. Sheikh, "Hybrid classifier based life cycle stages analysis for Malaria-infected erythrocyte using thin blood smear images," *Neural Comput. Appl.*, vol. 29, no. 8, pp. 217–235, Apr. 2018.

[159] H. S. Park, M. T. Rinehart, K. A. Walzer, J.-T. A. Chi, and A. Wax, "Automated Detection of P. falciparum Using Machine Learning Algorithms with Quantitative Phase Images of Unstained Cells," *PLoS One*, vol. 11, no. 9, p. e0163045, Sep. 2016.

[160] S. Widodo and P. Widyaningsih, "Software Development for Detecting Malaria Tropika on Blood Smears Image Using Support Vector Machine," *Int. J. Eng. Sci. Res. Technol.*, vol. 4, no.

REFERENCES

1, pp. 39–44, 2015.

[161] "The MaMic Image Database." [Online]. Available: http://fimm.webmicroscope.net/Research/Momic/mamic. [Accessed: 07-Apr-2019].

[162] M. A. Pourhoseingholi, M. Vahedi, and M. Rahimzadeh, "Sample size calculation in medical studies.," *Gastroenterol. Hepatol. from bed to bench*, vol. 6, no. 1, pp. 14–7, 2013.

[163] K. Suresh and S. Chandrashekara, "Sample size estimation and power analysis for clinical research studies," *J. Hum. Reprod. Sci.*, vol. 5, no. 1, p. 7, Jan. 2012.

[164] F. Fleuret, "Fast Binary Feature Selection with Conditional Mutual Information," *J. Mach. Learn. Res.*, vol. 5, pp. 1531–1555, 2004.

[165] N. Linder *et al.*, "A Malaria diagnostic tool based on computer vision screening and visualization of Plasmodium falciparum candidate areas in digitized blood smears," *PLoS One*, vol. 9, no. 8, p. e104855, Aug. 2014.

LIST OF RELEVANT PUBLICATIONS

Journals
1. **Sanjay Nag** and Ms. Nabanita Basu, "Critical Analysis of Malaria Parasite Detection Using Machine Learning Technique". Journal of Medical Imaging and Health Informatics, 9(4), 830–837. https://doi.org/10.1166/jmihi.2019.2677(**SCIE**)
2. **Sanjay Nag** and Prof. Samir Kumar Bandyopadhyay, "Malaria Disease Diagnosis- Current Status of CAD Based Approach, Cohesive J Microbiol infect Dis. 1(2). CJMI.000506. 2018.
3. Samir Kumar Bandyopadhyay, **Sanjay Nag**, Nabanita Basu, "Computer Aided Diagnosis Based Malarial Parasite Detection- A Review", International Journal of Experimental and Clinical Research, Vol 2017, Issue 01, pp 1-34, 2017.
4. **Sanjay Nag**, Roshni Dasgupta, Sayan Dutta, Dr. Indra Kanta Maitra, and Prof. Samir Kumar Bandyopadhyay, "Edge Detection of Digitized Histopathological Slide Images Using Dynamic Thresholding", European Journal of Pharmaceutical and Medical Research, Vol.3, PP.532-537, 2016.
5. **Mr. Sanjay Nag** and Prof. Samir Kumar Bandyopadhyay, "Identification of Malaria Infection Using HSV Colour Model and Dynamic Thresholding with Image Binarization", Innovative Journal of Medical and Health Science, Vol. 6, PP. 31 – 34, 2016.
6. **Sanjay Nag**, Satabdi Bhattacharya, Ranita Banerjee, Dr. Indra Kanta Maitra and Prof. Samir Kumar Bandyopadhyay, "Pseudo-Grayscaling Technique: A Pre-Processing Step towards Pathological Slide Analysis", European Journal of Biomedical and Pharmaceutical Sciences, Vol. 3, PP.385-390, 2016.
7. **Sanjay Nag.**, Nabanita Basu and Prof. Samir Kumar Bandyopadhyay, "White Blood Cell Segmentation and Malaria", International Journal of Current Medical and Pharmaceutical Research, Vol. 2, PP.202-206, 2016.
8. Samir K. Bandyopadhyay and **Sanjay Nag**, "Detection of Contour of the element of the WBC for Malaria Detection"- Advances in Computational Sciences and Technology, Volume 8, Number 1 (2015) pp. 7-15, 2015
9. Indra Kanta Maitra, **Sanjay Nag**, Pradip Saha and Samir K. Bandyopadhyay, "A Tree-based Approach Towards Edge Detection of Medical Image using MDT", International Journal of Computer Graphics Vol. 6, No.1 (2015), pp.37-56, 2015
10. Sangita Bhattacharjee, Jashojit Mukherjee, **Sanjay Nag**, Indra Kanta Maitra and Samir K. Bandyopadhyay, "Review on Histopathological Slide Analysis using Digital Microscopy", International Journal of Advanced Science and Technology, Vol.62, pp.65-96, 2014.

Book Chapter
1. **Sanjay Nag**, Nabanita Basu, and Samir Kumar Bandyopadhyay, Hybrid Approach towards Malaria Parasites Detection from Thin Blood Smear Image, Chapter 5, pp 93-122, Hybrid Intelligent Techniques for Pattern Analysis and Understanding, Ed. Bhattacharyya, S., Mukherjee, A., Pan, I., Dutta, P., and Bhaumik, A. K., Taylor and Francis, CRC Press, 2017.

CPSIA information can be obtained
at www.ICGtesting.com
Printed in the USA
LVHW081255210722
723979LV00012B/255